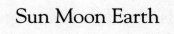

Sun Moon Earth

Also by Tyler Nordgren

Night Sky: A Guide to Our Galaxy

*Stars Above, Earth Below: A Guide to
Astronomy in the National Parks*

Sun

Moon

Earth

The History of Solar Eclipses
from Omens of Doom to
Einstein and Exoplanets

Tyler
Nordgren

BASIC BOOKS

New York

Books published by Basic Books are available at special discounts for bulk
purchases in the United States by corporations, institutions, and other
organizations. For more information, please contact the Special Markets
Department at Perseus Books, 2300 Chestnut Street, Suite 200, Philadelphia, PA 19103, or call (800) 810-4145, ext. 5000, or e-mail special.markets
@perseusbooks.com.

Designed by Jack Lenzo

Library of Congress Cataloging-in-Publication Data

Names: Nordgren, Tyler E. (Tyler Eugene), 1969–
Title: Sun, Moon, Earth : the history of solar eclipses, from omens of doom
    to Einstein and exoplanets / Tyler Nordgren.
Other titles: Solar eclipses
Description: New York : Basic Books, [2016] | Includes bibliographical references and index.
Identifiers: LCCN 2016013888| ISBN 9780465060924 (hardcover) | ISBN
    9780465096466 (ebook)
Subjects: LCSH: Solar eclipses.
Classification: LCC QB541 .N67 2016 | DDC 523.7/809—dc23 LC record
    available at https://lccn.loc.gov/2016013888

10 9 8 7 6 5 4 3 2 1

This book is dedicated to my father, who still feels terrible about me missing the 1979 eclipse. Don't worry anymore; it set me on the path to be the right person at the right place and time for 2017.

Contents

Prologue
xi

INTRODUCTION
From Omen to Awe
1

CHAPTER 1
A Day with Two Dawns and Midnight at Noon
7

CHAPTER 2
Two Worlds One Sun
29

CHAPTER 3
Shadows Across a Sea of Stars
57

CHAPTER 4
As Below, So Above
85

CHAPTER 5

The Eclipse That Changed the World

III

CHAPTER 6

Saros Siblings

143

CHAPTER 7

The Great American Eclipse and Beyond

171

CHAPTER 8

The Last Total Eclipse

191

Acknowledgments

209

Solar Eclipse Resources

211

Notes

213

Index

227

It is a spectacle pure and simple, the most magnificent free show that nature presents to man. . . . [N]ot to view the coming one would be literally to lose the opportunity of a lifetime.

—On the Solar Eclipse of 1925,
The New York Times

Prologue

I have spent my whole life watching the sky. As an astronomer, I've been to observatories all over the world and used every kind of telescope to look at distant star clusters and massive galaxies. Every clear night is an opportunity to experience something amazing. I have seen comets stretch across the sky, viewed sunlight glinting off the dust that floats between the planets, and witnessed a Milky Way so bright that the glow of its billion stars cast a shadow at my feet. But in all my life I have never seen anything as awe inspiring, as *awesome*—in the original definition of the word—as a total eclipse of the Sun. It is the only astronomical wonder that requires no telescope or complicated equipment to see. In fact, it looks even more spectacular to the eye than through the lens of any camera.

For an event that has at some point touched almost every place on Earth, remarkably few people have ever seen a total solar eclipse. The fact that *anyone* is able to see one is due to the great coincidence that our Moon is exactly the right size and distance from the Earth to cover the Sun completely. More often than not, the alignment between the worlds is imperfect, resulting in a partial eclipse where the Moon only blocks a portion of the Sun. At those moments, the Sun is still blindingly bright (literally) and so we are warned to use those little paper eclipse glasses that reward us with the strange sight of a Sun that is not fully there. It's likely that you have probably seen a partial eclipse without making any special effort.

But on those much rarer occasions, when the alignment of Sun and Moon is perfect, and you stand fully within the shadow of the Moon, you'll see a *total* eclipse. The shadow on the landscape is small—maybe no more than a few dozen miles wide—yet the motion of the Moon draws this darkness eastward for thousands of miles across our planet. This is the path of totality. For anyone on the ground, the experience can be either awe-inspiring or merely interesting, depending entirely on whether you are inside or outside that ribbon of darkness. Outside totality's path and the Sun is still blinding—use of those glasses is imperative for your safety. But stand inside the path and the temperature drops, birds grow quiet, shadows sharpen, and colors become muted and fade. Then, all at once, the Sun turns black and the stars come out. Overhead a ghostly aura streams outward around the Sun's dark disk.

Make no mistake, the difference between whether you're inside the path of totality or outside it is literally the difference between day and night. No other experience comes close to the multisensory strangeness of this most unnatural of natural events. From someone who has been there, trust me, the minute it's over, you'll wonder where and when to go to see another.

Totality changes everything.

It got to be pitch dark, at last, and the multitude groaned with horror to feel the cold uncanny night breezes fan through the place and see the stars come out and twinkle in the sky. At last the eclipse was total, and I was very glad of it, but everybody else was in misery. . . . Then I lifted up my hands— stood just so a moment—then I said, with the most awful solemnity: "Let the enchantment dissolve and pass harmless away!"

—MARK TWAIN, A CONNECTICUT YANKEE
IN KING ARTHUR'S COURT, 1889

From Omen to Awe

W hat if Christopher Columbus had been killed in the Caribbean? At one moment in history, on the night of February 29, 1504, the fate of the world we now know depended on the alignment of the Sun, Moon, and Earth in the form of an eclipse.

Columbus was in the midst of his fourth and final voyage to the New World when his aging ships became wrecked on the north shore of Jamaica. Initially the castaways found succor from the locals, but as the days turned to months their appetites outstripped everything available. Facing days on end without food, Columbus's crew mutinied and set upon the Jamaicans, who immediately rose in rebellion.

Whatever his gifts as a navigator, Columbus was a terrible ruler of men; he had already been fired as governor of the Indies for his incompetence and cruelty. Fearing for his life, he consulted the astronomical almanacs he used for navigation. In them he found that three days later, on the evening of the 29th, there would be a total lunar eclipse when the Moon passed through the shadow of the Earth. At those moments, the Moon takes on the reddish color of sunlight filtering through the Earth's atmosphere. Astrologers call this a "blood moon" for obvious reasons.

Assembling the Jamaican chieftains, Columbus told them that God was angry at their rebellion and would make His displeasure known by causing the Moon to be "inflamed with wrath." That night, when the Moon finally rose, a dark shadow had already begun to spread across its face, and the assembled islanders looked on in horror as the darkness spread. Eventually the Moon became the dark coppery color of blood.

Make it stop, the chieftains pleaded. To which Columbus replied that he'd need to retire to his cabin to pray on their behalf. In reality, he went there to keep watch on his hourglass, as his almanac had also revealed that it would take forty-eight minutes for the Moon to move through the darkest part of the Earth's shadow. When at last the sand ran out, Columbus stepped outside and proclaimed that God had answered his prayers. He would forgive their rebellion, provided they once more brought food for his men. In the midst of the Jamaicans' relief, totality ended, and the dark red blemish drained from the face of the Moon.

Almost four hundred years later, Mark Twain made use of this event for the purposes of a story in which a time-traveling Connecticut Yankee foretells a total solar eclipse, when the moon passes between Earth and Sun, to scare the medieval knights of King Arthur's court and save himself from execution. For the majority of human history, eclipses have been terrible apparitions. For people lacking knowledge of astronomy, they occur without explanation or warning. Many cultures throughout history have therefore constructed complex myths and rituals to explain why they happen to bring sense to the senseless, and to describe how those who are deserving can avoid the doom they seem to portend. From careful observation, however, we have discovered that these events are not random. In fact, they repeat in predictable cycles, and over time we have used their appearance to measure our world and reveal the mysteries of the universe.

Columbus's fate on that night in 1504 lay at the intersection of fear and calculation, mysticism and science, all of which are at the heart of how we humans experience eclipses. The fact that history balanced that night on the alignment of worlds (and the calculations of a man who had no idea where he really was) makes it all the more remarkable that today people travel the planet to experience those few fleeting minutes of totality, and when it is over, wonder when they can see another. Eclipses have made the transition from omens of doom to sought-after moments of awe. This is that story.

Skoll the wolf who shall scare the Moon
Till he flies to the Wood-of-Woe:
Hati the wolf. Hridvitnir's kin,
Who shall pursue the Sun.

—THE ELDER EDDA, GRÍMNISMÁL

CHAPTER 1

A Day with Two Dawns
and Midnight at Noon

It's not even lunch yet when something takes a bite out of the Sun. It's only a tiny notch at first, all but invisible without my cardboard eclipse glasses. Were it not for the shouts from the crowd around me on this August day, I never would have noticed. But now that I'm watching, I can see the bite grow bigger. It is the edge of our unseen Moon. The Sun is being eclipsed.

It takes forty minutes for the dark notch to grow so big that the Sun is now a crescent. But even with most of the Sun covered, the heat of the day is still intense. I take off my glasses and take refuge under a tree. There in its shadow I see a thousand bright crescents swaying in the grass to the time of the breeze in the treetop. Every one of them is an image of the Sun, projected on the ground when each tiny gap in the leaves overhead acts as a *camera obscura*, a pinhole camera. Nearby children have spotted them too, and begin to yell and giggle as they point and play among the bright little arcs. Had I not known what was happening before, this oddity would certainly have revealed the eclipse in progress above.

After an hour has passed, only twenty minutes remain until totality begins. The life-giving nature of the Sun is no longer an abstract concept: the sky has grown darker and colors are strangely wrong. The landscape is sapped of saturation. The worlds are aligning.

With ten minutes left, the conditions change quickly. The world has turned to twilight. The shadows of trees sharpen, as if lit by a single spotlight. Instead of coming from a round yellow disk set amid a bright blue sky, all illumination now comes from a narrow white crescent in a colorless dark vault.

I put my glasses back on for these final moments of the partial phase and can see the remaining crescent shrink as I watch. The crowd rises. Conversations hush, and I notice for the first time that all birdsong has ceased; the birds have returned to their nests to sleep in the unexpected night. An unseasonably cool wind blows across my arm as the temperature drops. The eclipse becomes a multisensory experience of sight, sound, and touch. So little of the Sun is left that surely totality should begin at any second, but I can't tear my eyes away to look at my watch. Even the passage of time seems affected. These last few seconds seem to expand rather than diminish.

Suddenly, the Sun's thin sickle of light breaks apart into an array of brilliant specks that dance and shimmer along the Moon's jet-black rim. They are called Baily's beads—the last rays of the vanishing Sun streaming through actual mountain valleys along the curved lunar surface. I finally remove my protective glasses to see them quickly wink away until a glorious diamond ring appears—a single glistening star set in a band of white radiance encircling the Moon.

Then the spot collapses upon itself and is gone.

We have reached totality.

Where before there was light and heat, now there is only a cold, black hole in the sky surrounded by a ghostly crown. The corona, a ring of immense pearly tendrils, envelopes the darkness and stretches off into the sky in all directions. It is unimaginably beautiful and only ever visible during these few precious minutes of totality. All around it are the brighter stars and planets, invisible until now; it is a day that has become night at noon, with the Sun, Moon, planets, and stars overhead all at once.

As an astronomer I know the mechanics of this celestial alignment, yet in this moment of totality I fully understand the difference between knowledge and feeling. The hair is raised on the back of my neck and my mind screams at the wrongness of what I am seeing. It is no coincidence that cultures from all over the world witnessed this sight with some degree of dismay. The Greek origin of the word "eclipse" is *ekleipsis*, meaning omission or abandonment. Ancient Chinese eclipse accounts contain the characters for "ugly" and "abnormal." For the Aztec, the eclipsed Sun "faltered" and became "restless" and "troubled."

These reactions make perfect sense when you consider that the Sun is the giver of life. When the Sun goes away without warning, it leaves behind the fear that it—our life source—might not come back. It is clear to me now why people throughout time did what they did to scare away the demons, chase away the jaguars, and slay the monsters they imagined devouring the Sun. The French astronomer and historian Jean-Pierre Verdet has found that this fear-fueled call to action was universal.

In Paraguay and Argentina, the roar of the crowds and barking dogs frightened the celestial jaguar that ate the Sun. In Scandinavia, Norsemen yelled to frighten away the demon dogs that the god Loki had sent to hunt and feed upon the Sun and Moon. The Ojibwe of North America sought to help the beleaguered Sun by firing flaming arrows to help him regain his light. In India, the people banged pots and pans to frighten Rahu, an immortal head who chased and ate both Sun and Moon. If they were loud enough, Rahu, startled, would drop the Sun from his jaws: totality would be averted and the eclipse would be only partial. For the Aztec, however, matters were more serious: "The common folk raised a cry, lifting their voices, making a great din, calling out, shrieking. . . . People of light complexion were slain [as sacrifices]; captives were killed."

Fortunately for any fair-skinned Aztecs, multiple total solar eclipses for any one location are rare. Though eclipses happen roughly twice each year, each one follows a different path across our planet. The patterns repeat in shape

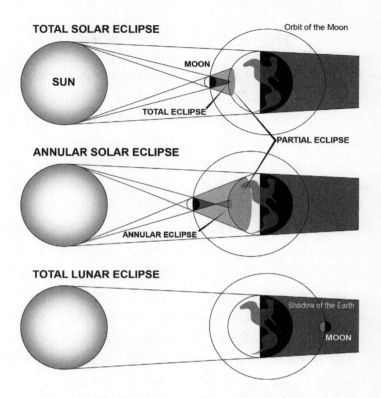

FIGURE 1.1. The three types of eclipses: solar eclipses occur when the Moon passes between Earth and Sun, while lunar eclipses occur when the Moon enters the Earth's shadow. Annular solar eclipses occur when the Moon is too far away from the Earth to fully cover the Sun, leaving a single "ring of fire" in the sky. (Image by the author)

every eighteen years, but each time, the path is positioned one-third of the way around the planet and a little farther north or south than before. As seen from a location high above the globe, the paths slowly spiral around the planet from pole to pole until eventually any spot on Earth can

expect, on average, to see totality every 375 years. Occasionally, the different paths do cross, and so every once in
a while, the rare person in a particular spot may live long
enough to see multiple solar eclipses over several decades.*
For cultures that looked to the sky for omens—where every
new star, comet, or eclipse could be the sign of the end-
times—imagine what seeing two total eclipses in one lifetime would have meant.

A thousand years ago, in what would become the
American Southwest, Chaco Canyon was the ceremonial
center of the Ancient Puebloan people (whom we used to
call the Anasazi). There, on the side of a massive boulder,
is a pictograph unlike any other. It features a large circle
pecked into the yellow sandstone surrounded by strange
looping tendrils similar in appearance to those of the solar
corona. It is thought to be a record of the total solar eclipse
of July 11, 1097, one of *three* solar eclipses (two of them
total) visible there over a period of fifty-eight years at the
height of their culture. To the upper left is a second small
circle, precisely where Venus would have appeared at the
moment of totality so long ago.

Imagine the effect such an apparition would have had
for a Sun-watching people at the heart of their ceremonial
society during a decade of extreme drought when the climate was changing for the worse. For a people in the midst

---

* Though no total solar eclipse has touched the mainland United States between 1979 and 2017, the residents of Carbondale, Illinois, will be lucky
enough to see two total eclipses in seven years: first in 2017, and again in 2024.

of extreme cultural and environmental crises, might such an eclipse have been yet another contributing factor in what made the Puebloans, also called the Chacoans, eventually wall up their monumental "Great Houses," set them aflame, and abandon their canyon?

Even today, eclipses play on our fears. The American anthropologist Ward Keeler described the event of June 11, 1983, when a total solar eclipse swept across Indonesia.

> The air became very still and Java's lush vegetation glowed in the eerie light characteristic of sunset in the tropics. As at sunset, too, the horizon turned red, but it did so not only in the west but in all directions, and in the half-light distant volcanoes usually obscured by the glare of the Sun became visible. For the four minutes of total eclipse, the Sun, almost directly overhead, looked like a black ball surrounded by a brilliant white light. Most eerily of all, in one of the most densely populated rural areas in the world, there was no traffic on the roads, no movement in towns or villages, and no one watching the eclipse.

For weeks prior to the event, newspaper reporters, radio announcers, and TV stations had gone to great lengths to warn people about the event for fear that they would damage their eyes peering up at the vanishing Sun. Posters were prominently displayed in villages across the country bearing the message that watching the eclipse would cause viewers to go blind. The warnings were so effective that residents stayed in their homes during the eclipse. They dared not even look outside for fear of the Sun's "sharp rays."

I know that mistaken fear firsthand. The last total solar eclipse to touch the continental United States did so in Portland, Oregon, on February 26, 1979. I was a boy, only nine years old then. In my fourth-grade class, we made clay medallions of the upcoming eclipse. While others painted black circles with yellow crescents in representation of the partial phase, I had found library books showing the corona, and so I carefully painted the billowing white ring around the central black hole. On the morning of the eclipse, my school canceled classes. Yet rather than go out and see the sight for myself, I hid indoors with the curtains drawn. Local TV and radio stations had been inundated with the exact same messages of fear that would later be broadcast all over Indonesia. I hid indoors, terrified of the same mysterious rays with the power to make me go blind if I so much as got a glimpse of the eclipsed Sun.

Today I know that there are no special rays, sharp or otherwise. The Sun is just as bright on any ordinary average day as it is on the day of an eclipse. During the time that the Sun is partially covered, it is still bright enough that staring at it for even a couple of seconds can cause permanent damage to the retina (just like on any other day). For this reason, eclipse glasses are necessary when the eclipse is partial; once totality begins and the Sun disappears, you can take your glasses off, and the spectacle is as safe as it is awesome.

Yet, in our zeal to be "safe," we flood the airwaves with our fears, never with our hopes. That is why, to this day, my first eclipse memory is of watching the events unfold on my RCA color TV (snapping photos off the screen with my plastic drugstore camera). My only direct experience of the event itself was noticing how dark the house became as totality passed unseen overhead. It would be thirty-eight years before a total solar eclipse would touch this country again, and I have spent every one of those years wishing I'd turned around, gone to the window, parted the curtains, and simply looked up.

My career as an astronomer has taken me around the world since then, partly in pursuit of exactly that which I so narrowly missed when I was nine. Yet though I have seen multiple solar eclipses since, I will never be able to see the one I missed that day. Every eclipse is different. The shape of the corona, the streamers and jets that are such a startling phenomenon of totality, is dependent on the conditions taking place on the Sun at just that moment, and its exact shape is unknown until the instant of totality.

The search for meaning in celestial events is the purview of astrology. Astrological records of ancient China claimed that solar eclipses were a reflection of the quality of the king, and the corona's appearance was said to reveal the political plots at work behind the throne: "[If the king] does not share his fortune with his subjects, the condition is called unstable. Then there will be a total eclipse with Sun being black and its light shooting outward. . . . If there

are two ear-rings beside the Sun during eclipse while in the
east, west, south, and north corners there are white clouds
shooting outward, then the whole country will be in war."
But China is not the only place where the sky was searched
for meaning: court astrologers all over the world have done
the same for millennia. A comet appears in the sky? The
king will be overthrown. A new star (a supernova) appears
in the constellation of Leo? A new king will be born. The
Sun is eclipsed? The king is wicked. As a steely-eyed scientist who prides myself on my reason and yet is moved to awe
by such a rare and beautiful phenomenon, I can understand
the ancients' desire to associate eclipses with events of great
importance.

If eclipses were thought to mark momentous events,
then the reverse was also true: certain events that the original chroniclers wished to appear to be momentous were
made so by adding an eclipse, even if no actual eclipse was
handy. The end-times stories of Ragnarok and the Rapture
in the Norse and Christian traditions, respectively (as well
as the Crucifixion in the gospels), are accompanied by the
Sun turning black. These details have often been interpreted as references to total eclipses, even if the description
of the event, as in the case of the Crucifixion, bears little
similarity to an actual eclipse.

Even in modern times, momentous events seem to have
their associated eclipses. To anyone living in Boston, Massachusetts, the night the Red Sox broke their eighty-five-year-
long World Series "curse" is as momentous as they come. It

was made all the more so by the fact that it only happened during the final moments of the total lunar eclipse of October 27, 2004. At that instant, the Moon became red, as it was lit by the light of every sunrise and sunset happening on Earth at that moment. Perhaps the Red Sox could only win by the light of a red Moon?

For such an awe-inspiring sight, throughout most of history the ability to call it into existence was a sign of one's power with the gods. A little over three hundred years after Columbus scared the Jamaican chieftains with his calculation of a lunar eclipse, another European thought he'd try the same trick with the Plains Indians in the American West. There, a local doctor in a small town in the Dakota Territory in 1869 read in his almanac that there was about to be a total eclipse of the Sun. Eager to impress upon the local Sioux the power of the White Man's magic and healing arts, he told them the exact date and time the eclipse would begin. And it would last, he said, until he saw fit to stop it.

When totality occurred, the Sioux, rather than cowering in fear, raised their rifles and fired into the air. When the Sun came out again, they calmly stated that "the doctor could predict the eclipse, but they could drive it away."

That eclipses can be predicted years in advance and all over the globe—and that people made these predictions for over a thousand years without computers—is truly remarkable. Wish to see a total solar eclipse? Astronomers can now tell you the location and time of any future eclipse, down to the mile and the second. More importantly, the proof of whether or not we are correct will be waiting for you when you get there: either you see the corona or you don't. If you don't, then we learn we didn't understand the world as well as we thought, and we will seek to correct what we failed to get right. This is the power of science and the process by which we have learned everything we know about the physical universe in which we live.

Astrology, like astronomy, makes predictions. Astrologers claim that the position of the Sun, Moon, and planets at the time of your birth influences your personality and fate. It tries to identify auspicious dates, opportune investments, and compatible mates. The one thing it does not do, however, is reevaluate its assumptions when it's wrong. Only science does that. Yet in a 2014 National Science Foundation survey, nearly half of all Americans (45 percent) responded that they believed there was some scientific basis to astrology. Imagine my disquiet when, during my most recent trip to the doctor, the nurse drawing my blood looked at my paperwork and said with a smile, "Oh hey, you're a Scorpio too!"

The primal appeal of pseudo-sciences like astrology is understandable. Life is full of dangers and misfortune that plague us at random. Astrology gives us hope that there is a cosmic reason, a connection with the Sun, Moon, and stars that gives order to the apparent chaos we encounter. These emotional needs seem all the more necessary when the heavens behave in ways that aren't normal, as in an eclipse.

Yet the science of astronomy reveals a far more direct way in which the heavens guide our lives on a daily basis. The Sun gives us light, heat, and food. Those organisms that don't feed directly on sunlight feed on those organisms that do. Our everyday concepts of position, direction, and time intimately depend upon the motion of that Sun. What is a "day" but a single rotation of our planet? A "year" measures its orbital motion about the Sun while the orbit of the Moon marks the period of time we call a "month." Imagine every task, chore, rite, or celebration that happens on an annual basis, and you will understand why we needed astronomers in our past. Could civilization have arisen without astronomy? Might we all be the metaphorical descendants of astronomers?

Who were the first astronomers? To answer that question, we must imagine a family tree of our distant ancestors. Four million years ago, our small *Australopithecus* ancestors first began standing up on the African savanna. As the American astronomer Neil deGrasse Tyson has said, "Once we were standing upright, our eyes were no longer fixated on the ground." Out in the open, away from the cover of trees, these early ancestors lived under a night sky more vivid than the one we can now see from almost anywhere on Earth. We don't know if the australopiths noticed, but we do know we are not the only beings on the planet today who notice the sky. Sea turtles, birds, and dung beetles all make use of the stars and the Milky Way in navigation—but we wouldn't call them astronomers. So our *Australopithecus* ancestors were not the first astronomers, even if they did something similar. Use alone isn't science.

By 2.5 million years ago, our *Homo habilis* ancestors were following animal herds in their annual migrations, and evidence exists for seasonal camps during their travels. Did they plan them by noting the passing of the seasons with the changing Sun and stars in the sky, or did they merely set up new camps as they kept close to the animals they were following? Lions, like our *H. habilis* ancestors, follow herds, but they aren't scientists. So perhaps astronomy didn't begin here, either.

A million years later, our *Homo erectus* forebears mastered fire, which for the first time extended the day's work into darkness. Perhaps the first constellations were imagined during those nights, but if so, we have no record of them.

Only 60,000 to 100,000 years ago, the first *Homo sapiens* fed on shellfish from tide pools on the south coast of Africa. The tides were tied to the Moon and changed each day, both in time and size, as the Moon went through its phases. That's still the case, of course. And for the early members of *H. sapiens*, there would have been a benefit to understanding these patterns: those who did would have been able to feed themselves and their families a more varied diet than those who did not. So one might conclude that this is where astronomy began.

But consider for a moment what is required to make the mental connections needed for astronomy. The ocean tide is a direct physical effect; it gets you wet when it rises and reveals its food when it recedes. The Moon, by contrast, is so far away you can't touch it, hear it, or smell it. There's no reason, intuitively, that the tides and the Moon should be connected, and what connection there is can only be revealed through observations over a long period of time, requiring memory, abstract pattern recognition, and a belief in an underlying order or relationship.

The archaeologist Steven Mithen referred to these skills as "cognitive fluidity": the ability to synthesize different forms of intelligence (such as knowing how to build fires, make tools and weapons, and interact in a group and structured society) and to combine these in ways that incorporate abstract ideas, myths, and long-term observations. Evidence for this fluidity appears only after about 60,000 years ago. It is evidenced in the first examples of representational art and in some bone artifacts. If this is the earliest time when something resembling science could have arisen, then the first, most unambiguous evidence of human astronomical knowledge should be more recent still.

Less than a mile from the Nile River, in what was once ancient Nubia, there is a complex of graves. Here, in 1964, archaeologists discovered fifty-eight ancient skeletons all buried on their left sides, facing south, with their heads to the east, toward the rising Sun. "The burial positions," an astronomer wrote in 2000, were "remarkably uniform," making it unlikely to be coincidence. The simple fact that the skeletons point eastward means that between 12,000 and

14,000 years ago, someone knew how to identify one of the four cardinal directions. These directions are defined by the sky. The east is where the Sun rises, the west where it sets. The line joining north to south is where the Sun is at its highest during the day, and at night (at least in the Northern Hemisphere), north is the direction around which all the stars turn. Here in the Nubian Desert is actual evidence of astronomical knowledge and of its association with some abstract, intangible meaning.

When one first asks the question, "Why does the Sun rise in the east?" there are two paths to follow for an answer. One path leads to science, the other to religion. For most of human history these paths ran side by side and were often indistinguishable. The ancient answer, "Because the gods make it so," covers a lot of phenomena and is difficult to refute. This is what we see in the stories of eclipses. Demons and deities eat the Sun and the Moon and do so for reasons known only to them.

This is a tricky path to follow, because any phenomenon we don't understand can always be blamed on the gods or a god. Why does the Sun rise in the east and not fall from the sky? It's the work of Apollo and his gleaming chariot. Why do the Sun, Moon, and stars circle overhead? Because God has placed the Earth at the center of the universe around which all things move. In recent years, this same kind of reasoning has been extended to biological evolution by those who believe the process is too complex to have occurred without an Intelligent Designer.

This explanation is called the "God of the Gaps," a term first coined by Henry Drummond, a nineteenth-century Scottish evangelist. Over time, as we discover more about our world, the gaps in our knowledge grow smaller, as does our need for miraculous intervention to explain what is seen. This is fair to neither science nor religion. For the religious-minded individual who looks for physical proof that God is at work in the cosmos, the duties of His job grow less consequential with every year. God gets demoted from being the Prime Mover of the heavenly firmament to merely twiddling the knobs on the values of a few physical parameters. It's unfair to science because once a miracle is invoked (in essence, to say what is unknown is unknowable), all further investigation stops. After the "Miracle Card" is played, there is no reason to keep testing hypotheses.

So while the question "why" can be fraught with metaphysical traps, science also asks "how"—a question that has answers open to direct experimentation. How long is a day? How can I learn this from the changing position of the Sun in the sky? How do the locations of the Sun and the stars at night define direction and the passage of seasons?

For careful observers, the sky becomes a calendar easily used to predict the changing seasons upon which individual and complex society's survival depends. It is, perhaps, then no coincidence that the first signs of agriculture are found at roughly the same period in time as the Nubian graves. To know when to harvest and plant, agriculture requires astronomers.

This transition from a chaotic world of seemingly random changes to a predictable world of returning stars, rain, and food is embodied in the story of eclipses. Like the ancient Chinese astrologers who courted palace intrigue, or even Christopher Columbus saving his own skin, whoever could understand the motion of the heavens, and thus predict an eclipse, had the power to declare why they occurred and impart order on chaos. In a modern world where the number of Americans who believe there is some science to astrology is twice that of the number who accept evolution, we are still in thrall to the cycles and patterns of the sky.

My attention returns to the black Sun overhead. A single needle of light bursts forth into a second diamond ring even more beautiful than the first. The light has returned too quickly, and the Sun is once more too bright to behold. It is over, and my first thought is, "When can I see another?"

That I chase eclipses where my ancestors feared them is not to say that they were foolish to fear the sky. Thanks to science, while we no longer blame demons and believe in omens, we do understand that ancient terrors like comet impacts and nearby supernova explosions could kill most life on Earth (and in some instances may already have). And while eclipses themselves are without danger, how we react to them could be if we fail to take care of rudimentary eye protection. Far more importantly, eclipses remind us of how dependent we are upon the single star that is our Sun.

Over the past three millennia, eclipses have made the transition from terrifying omen to scientific tool to benign tourist attraction. In this book, we will follow that story, from the shamans and astrologers who divined the patterns of eclipses and perfected their predictions to the philosophers and scientists who discovered the true cause of eclipses and used them to measure the world and explore the universe beyond. Eclipses, on this world and others, now reveal that we are just one planet in an ever-growing family of planets throughout the galaxy in an ever-expanding universe. This is the story of science and ourselves, down the path to which totality leads.

*The Eclipse of the Moon and Sun is a Thing through-
out the universal Contemplation of Nature most
marvelous, and resembling a Prodigy, and shews the
Magnitude and Shadow of these two Planets.*

—PLINY, HISTORY OF NATURE, BOOK II, CHAPTER X

CHAPTER 2

Two Worlds One Sun

The Sun shines on Mars as it does on Earth. For at least a billion years, only dusty red rocks have cast shadows on the fourth planet from the Sun. Lately, however, the Sun has been lighting up something new on Mars, and it's casting a shadow that slowly changes with the hours. It is a sundial.

In fact, there are now three sundials on Mars. Since 2004, each of the rovers sent to the planet by the National Aeronautics and Space Administration (NASA) has possessed a sundial—also known as a MarsDial—that I and a small team of other artists and astronomers designed to cast shadows across a plate of concentric gray circles and four colored squares. Scientists on Earth use the colors to calibrate how our robotic explorers "see." Periodically, the cameras of the rover are pointed where the Sun should appear. If the Sun's disk is perfectly centered in the photos, then the rover must be located and oriented exactly as thought. It's a tiny example of the scientific method in action: a test of our ideas versus reality.

A sundial on a nuclear spacecraft, navigating by the Martian sky, may seem like an odd combination of old and new (and an equally strange way to begin a story of eclipses), but in a way, it's not odd at all. We have depended on the Sun and its motion for as long as we have been human, so it's only natural that we should use whatever it can tell us no matter what planet we happen to be on (at least within our own solar system).

On Earth and Mars, the motion of the Sun in the sky defines the cardinal directions and the lengths of both day and year. A Martian day (called a *Sol*) is only forty minutes longer than our own, while a Martian year is 88 percent longer, owing to its slower speed around the Sun and the greater distance to travel. Yet each planet's axis is tilted, so Mars experiences its seasons just as we do on Earth, though those on Mars last comparably longer. Only our month, defined by the approximate time it takes our Moon to go through its phases, has no reasonable analog on the Red Planet, as that world has two moons, each with a different orbital period.

For thousands of years, on Earth at least, the proof that the calendars we made kept accurate track of these days, months, seasons, and years was that they were able to stay synchronized with the celestial cycles. Eclipses were a benchmark for checking whether we were right. Because eclipses are caused by the shadows the Sun casts on both Earth and Moon, eclipses should happen only when all three worlds line up. Lunar eclipses occur only at Full

Moon, when Moon and Sun are on opposite sides of the Earth, visible to everyone on the shadowed side of Earth. A solar eclipse occurs two weeks (or half an orbit) later, when the Moon passes between Sun and Earth and its shadow falls somewhere on our planet. Only those fortunate enough to stand in the relatively small shadow get to see a solar eclipse.

Since it takes the Moon a month to orbit the Earth, it stands to reason that every month should see a lunar eclipse at Full Moon, and two weeks later a solar eclipse when the Moon is new. How I wish this were so. Sadly, the Moon's orbit is tilted by five degrees to our own around the Sun. For the majority of New Moons, the Moon passes without notice slightly above or beneath the Sun, and two weeks later similarly misses the shadow that our planet projects out into space. Only when the alignment is exactly right do eclipses occur.

Let's see how a person, without spacecraft or computers, without knowledge of planetary orbits or even of planets, would have figured this out. Science begins with observation. For most of human history, our lives depended on observing the cycles of the Sun. By noting where it rose each morning, our ancestors could measure the changing of the seasons, and thus determine when to hunt, plant, and harvest. The axial tilt of our planet is what gives us our seasons, and even a simple sundial, including those on Mars, reveals their passage. As winter turns to summer, the Sun rises farther to the north and climbs higher in the sky. Noontime shadows grow shorter each day, and then, as summer gives way to winter, the low southern Sun casts longer shadows once more.*

All over the Earth there are ancient buildings, petroglyphs, and geographic alignments that our ancestors used to mark the passage of light and shadow over days, months, and years. The most famous of these is Stonehenge in southern England. Two concentric rings of massive standing stones with horizontal slabs stretched across their tops form its most prominent feature. These date to at least 2300 BCE,† though other surrounding features are older. Radiocarbon dating of burnt wood reveals that people have been in the area since as far back as the eighth millennium BCE.

Outside these rings stands a large pointed boulder called the Heel Stone. Archaeological evidence indicates that a similar boulder was once set into the ground beside it. From the center of the stone rings toward the point between the Heel Stone and its missing twin, an observer 4,300 years ago would be looking precisely at the rising Sun at its northernmost point on the horizon on the longest day of the year: the summer solstice.

I have personally seen a similar alignment of Sun and structure much closer to home in Southern California. There, on the spring and fall equinoxes, the last rays of the

---

* Mars, by chance, currently has the same axial tilt as the Earth and so its seasons are similar to Earth's, though more extreme due to the size and shape of its orbit.

† BCE means Before the Common Era, equivalent to BC.

setting Sun shine down a long passage to a doorway into which a crystal has been set. Through it the Sun's light is refracted in a rainbow of color across a small vestibule and through a large room where it shines directly onto a square niche imbedded into a wall. It's a magical sight that I have been lucky to see every spring and fall for the past decade— the fact that it's also the living room in my own house in no way diminishes its power to delight me. I live in my own personal Stonehenge, by accident rather than by design, and in a country where many city streets were laid out by surveyors precisely along the cardinal directions, there is an excellent chance that you do, too.

One such alignment occurs in the heart of New York City, where the streets are arranged in a grid angled 29 degrees east of north. Twenty-one days before and after the summer solstice (typically on or around May 28 and July 12), the setting Sun shines perfectly down the canyon of streets on the island of Manhattan. It is a phenomenon the astronomer Neil deGrasse Tyson calls Manhattanhenge, and there are actually people who will brave evening rush-hour commuters to dash out in traffic and photograph this alignment of city and Sun.

Whether Stonehenge was initially planned as an observatory, was an observatory by accident like the one in my house and New York City, or was never an observatory at all, the fact remains that it works as one today, and so, too, marks the movement of the Moon. It takes the Moon 29.5 days to complete one *lunation* (the time between two

Full Moons*). One of Stonehenge's stone rings features 30 archways, one of which is half as wide as all the rest. Outside this ring are two more rings of holes in the ground, one with 29, the other with 30. Is either of these symbolic of the Moon's 29.5-day lunation? With enough boulders and holes, almost any astronomical connection can be found, but it is intriguing nevertheless.

What is known with certainty is that cultures all over the world continue to mark time by the sunrise, including the Hopi in the American Southwest, whose ancestors, the Chacoans, built the "Great Houses" mentioned earlier. The alignment of the Great Houses appears to keep track of the same celestial cycles as the standing stones at Stonehenge.

For those who do not live someplace where the changing location of sunrise is obvious, the Sun provides another clue: it moves against the background stars with the seasons (and thus we see different stars at night in summer and fall). The path the Sun follows across these stars defines the plane of the Earth's orbit and is called the *ecliptic*. It crosses twelve prominent constellations, and the Sun passes through approximately one of them every month. You've likely heard of these constellations: Libra, Scorpius, Sagittarius, and so on. Western astrology is based on the premise that your personality and fate are influenced by whichever one of these constellations the Sun was in front of on the day you were

---

* This is slightly longer than the Moon's orbit around the Earth relative to the stars (27.3 days), since the Earth's motion around the Sun causes the Moon to have to go a little bit farther each month to once more align with the Sun.

born. The prevalence of daily horoscopes says a lot about astrology's popularity, if nothing about its accuracy.

Because of the Moon's orbital tilt, the Moon spends half the month above the ecliptic and half below. Where the Moon crosses the ecliptic marks a *node* (Latin for "knot"). As viewed from the Earth, there are two nodes on either side of the celestial sphere that is our sky. Through one, the Moon passes upward through the ecliptic, and through the other it passes back down. When the Moon and the Sun both cross a node together, an eclipse takes place. Twice each year, the Sun is close to a node when the Moon goes

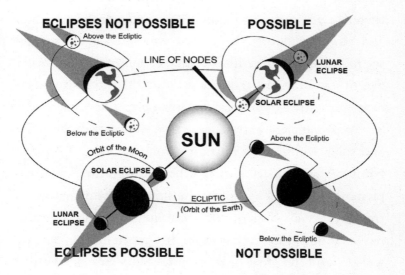

FIGURE 2.1. Because the plane of the Moon's orbit is tilted with respect to the orbit of the Earth around the Sun, eclipses are only possible when the "line of nodes," the point at which the Moon crosses the Earth's orbital plane, points toward the Sun. (Image by the author)

sailing by, and thus twice each year we see solar and lunar eclipses somewhere on Earth.

If the Sun and Moon were mere points of light, they would have to perfectly cross the node at precisely the same instant for an eclipse to occur. But the Moon and the Sun both span about half a degree on the sky. This produces a period of about 34 days, called an *eclipse season*, during which each body passes close enough to a node at Full or New Moon for an eclipse to take place, even if only partially.

The direction in which the Moon's orbit tilts slowly changes over an 18.6-year period. The positions of the nodes therefore drift westward along the ecliptic, and so the Sun and Moon encounter them slightly more often than every six months. Lunar eclipses therefore happen either every five or six lunations (every 148 or 177 days), with a solar eclipse occurring two weeks before or after each one.

We see ancient knowledge of this pairing between lunar and solar eclipses in a narrative from the Pomo people, a Native American tribe in Northern California. Their explanation of eclipses involves the story of a bear that walks the Milky Way. When he comes upon the Sun, the Sun refuses to step out of the way. For his impertinence, the two wrestle, and the great bear bites the Sun: "Sun got bit [by] bear" is the meaning of the word the Pomo used to describe an eclipse. After his battle with the Sun, the bear comes upon the Sun's sister, Moon. She, too, refuses to step aside, and again there is a great fight in the sky.

Back at Stonehenge, there are 56 *Aubrey holes* in a giant ring around the standing stones. The British astronomer Fred Hoyle hypothesized that these pits could be used to keep track of when lunar and solar eclipses occur. The ancient people of Stonehenge could have periodically moved a series of four stones from one hole to another at different rates, one for the Moon, one for the Sun, and two that were always kept a half circle apart to represent the nodes. An eclipse was possible when the Moon and Sun stones fell on either the same node stone (a solar eclipse) or opposite node stones (a lunar eclipse). With little more than these measures, a person could keep track of the phases of the Moon and the times of eclipses under Britain's cloudy skies and remain remarkably accurate for years.

The Mayans, too, were aware of eclipse seasons. One of the few remaining Mayan books, called the Dresden Codex after the city where it now resides, reveals multiple entries of the numbers 148 and 177 in the Mayan's unusual base-20 counting system of bars and dots. While it could be a coincidence that these are the exact numbers of days between lunar eclipses, the Mayan artist/author/astronomer who composed each page went on to make their meaning clear. At the center of each tally, the unknown scribe placed an elaborate half-black and half-white figure portraying the Sun and Moon, including one where it is about to be eaten by an enormous snake.

There are numbers other than the intervals between eclipse seasons that appear in ancient manuscripts and reveal the awareness of the frequency with which they occur. Consider that total solar eclipses can only happen at New Moon (every 29.53 days), during eclipse seasons (on average, every 173.3 days), and when the Moon is nearest to the Earth in its noncircular orbit (every 27.3 days). How often does it take for each of these cycles of different lengths to come around again like hands on a clock? Every 6,585.3 days (18 years, 11 days, and 8 hours), the Moon will fully eclipse the Sun at the same node, at the same time of year, and with the Sun and Moon in nearly the same constellations as it did before. For no shorter period of time do all of these three cycles of differing lengths coincide.* We call this the *Saros cycle*.

Eclipses separated by this amount of time are said to be of the same Saros, and each Saros is numbered. The first total solar eclipse I saw was of Saros 145 on August 11, 1999, in Hungary. The eclipse in the United States on August 21, 2017, is 18 years and 11 days later (if it were visible in Hungary it would be August 22) and therefore must be Saros 145 again.

The difference between the eclipse being visible in Europe in 1999 and being visible in North America 18 years and 11 days later is due entirely to the remaining 8 hours (0.3 days) in the Saros cycle. Between two eclipses of the same Saros, the extra 8 hours means the Earth will have rotated an extra third of its day, moving the eclipse a third of the way around the Earth to the west.

---

* This number, 6,585.3 days, is exactly 223 lunations, 38 eclipse seasons, and 239 orbits of the Moon.

After 3 cycles of 18 years (called an *Exeligmos*), totality once again occurs at the same longitude on Earth. But now 33 extra days have passed, and the Sun is a little higher (or lower) in the sky, depending upon the season, which shifts the path of totality a little farther north or south (depending on whether the Moon is ascending or descending through the node). Over time, total eclipses of a single Saros will begin at one pole of the Earth, slowly leave tracks spiraling around the planet, cross the equator, and after 1,300 years "retire" the Saros as it produces its last eclipse at the other pole. The solar eclipse of August 21, 2017, is the sixth total eclipse of Saros 145. There will be 35 more before it is done with us in 2143.

To notice cycles that repeat on a timescale of decades hints at cooperation over generations requiring careful recordkeeping. The first unambiguous record of the Saros cycle comes from the Chaldeans, who in 626 BCE ruled an empire that contained the ancient city of Babylon and extended from the Mediterranean eastward to the Persian Gulf and from the Red Sea northward into modern Turkey. What we know of their astronomy comes largely from the writing of later Greek astronomers and from a handful of small clay tablets full of closely spaced cuneiform markings. These tablets were used to record the names of kings and the dates of their reigns.

Perhaps as long ago as the eighth century BCE, the Chaldeans kept daily astronomical diaries recording the positions of the Moon and planets relative to the stars. Their observations were so precise that recent analysis of these tablets reveals they had developed a rudimentary form of calculus to keep track of the planet Jupiter's speed across the background stars. As far back as the seventh century BCE, these diaries also kept track of the weather as well as of economic and political events. I can imagine a Chaldean astrologer/astronomer making a note of everything he saw each day in hopes of identifying patterns that might indicate what phenomena bode well for the king and what occurrences bode ill.

One pattern becomes obvious in what are now referred to as the Saros Cycle Texts. These are multiple tables of kings' reigns and dates arranged in columns of 5 and 6 lunar months with columns precisely laid out 223 lunar months apart (223 months add up to 18 years). Though no mention is made of eclipses, modern calculations show almost perfect agreement with lunar eclipses visible in Babylon during that time. The fact that these dates were given with the names of known kings implies, of course, that the tablets were written after these eclipses would have come to pass.

What evidence is there that after noticing these cycles, ancient astronomers made the connections necessary to predict their future occurrence? Today we would refer to that as forming a hypothesis. If similar eclipses happen every 18 years, then once one sees the pattern, one should be able to predict the next.

The earliest account of someone predicting the exact time and place of a solar eclipse is from the Greek historian

Herodotus. In his *Histories*, written around 450 BCE, he told of a five-year-long war between the Lydians and the Medes that had taken place a hundred years earlier in what is now central Turkey:

> In this war they brought about a battle by night; and the engagement came about in the sixth year when they were still contending with each other at war on an equal basis, when it happened, as the battle was beginning, that day suddenly became night. (Thales the Milesian predicted to the Ionians this change [of day to night] would come about, setting beforehand the favorable period in which the ominous event did indeed happen.) When the Lydians and Medes saw it become night instead of day, they quit the battle and rather made haste on both sides that peace came about.

Historians and astronomers as early as Pliny the Elder in 77 CE have interpreted this "day turned to night" as a solar eclipse. Numerous attempts have been made to pinpoint exactly which eclipse this would have been. Unfortunately, the only solar eclipse that appears to match in both time and place (May 28, 585 BCE) was partial, and certainly not enough to have turned "day to night." Nevertheless, Thales could have been aware of the Saros cycle, and if he had heard of its previous occurrence over Egypt and the Persian Gulf on May 18, 603 BCE, then he could easily have predicted the day the next would occur.

As a result, there are astronomers who would like to see the site of the ancient battle declared a World Astronomical Heritage Site to help preserve its place in the history of science. At the very least, whether Thales actually predicted a solar eclipse or the story is a complete myth, the fact that writers of the time believed it was possible to predict an eclipse means that these amazing spectacles were no longer considered the work of demons or gods, but had been placed firmly in the realm of natural phenomena.

Predicting eclipses was one of the central roles of Chinese court astrologers. Like their colleagues in Babylon, the Chinese astrologers were responsible for recording eclipses, and they noted virtually every one that could be seen in their region over a period of more than 3,000 years. Accurate prediction and observation of heavenly events was necessary for effective time-keeping and ceremony. If an eclipse occurred as predicted, it meant you understood the periods of the Sun, Moon, and Earth, and thus your calendar of months and years accurately reflected the seasons. An accurate calendar meant you could successfully identify the dates of holy days important to unite your people, praise your rulers, and appease the gods.

This tradition continues today, not just in China, but also in most other areas of the world. Holidays like Christmas that occur on the same date every year are essentially tied to the period of the Sun (or, more accurately, the period of the Earth *around* the Sun). But each of the three main monotheistic religions has holidays that float from date to date because they are tied to the period of the Moon. Ramadan is the ninth month of the Islamic calendar, where each month begins at the first sighting of the crescent New Moon after sunset. Passover, like Ramadan, is also tied to the Moon. It occurs on the 15th day of the Jewish month of Nisan, which begins at dusk with the New Moon. Since 15 days is almost equal to half of the lunar cycle, Passover occurs at the Full Moon, and to keep it in the spring it has become the first Full Moon after the spring equinox. In Christian tradition, the Last Supper of Jesus was a Passover Seder, after which followed the crucifixion and resurrection. So Easter is now set as the first Sunday after the first Full Moon after the spring equinox.

In our modern world we may no longer worship the Sun and the Moon, but our worldwide religions are still tied intimately to their motions. At the very least, each year individuals celebrate life's victories and tragedies by the number of trips around the Sun they have made since their occurrence. Ours is a world enumerated, regimented, illuminated, and measured by astronomy.

But what do these millennia of observations, these innumerable cycles of Moon and Sun, traveling nodes, recurring eclipses, and repeating Saros reveal about our universe? If we derive our notions of direction and time from the Sun and stars, then what do the Sun and stars tell us about our place in space and time among them? And how do we know they are correct?

Stand in the Piazza San Marco at the heart of Venice, and you will see a giant model of the heavens in

beautiful gold and lapis blue counting the hours from the
Torre dell'Orologio. This spectacular clock dates from the
1490s and shows a golden Earth at the center of its massive
wheel. Around it spins the Moon, its half-blue and half-gold
sphere slowly rotating with the lunar phases over a period
of 29.5 days. Farther out, an ornate golden Sun glides across
the 12 constellations of the zodiac, in keeping with its actual position among the stars. Only in the very last ring are
24 giant roman numerals arrayed, revealing the time of day
by their alignment with the Sun. During the Middle Ages,
clocks like these could be found all over Europe, although
few as grand. For those who were no longer attuned to the
patterns of the sky, they served as a direct visual reminder
in the heart of each community that there was, astrologi-
cally speaking, a comforting order to life. The other thing
these clocks provided was a model of the universe that re-
assuringly places us at its center. Around us the Moon, Sun,
and planets (in that order) move against the background
stars. For the natural philosophers concerned with understanding the nature of reality (who would one day be
known as scientists), eclipses were a means to an end. They
were a tool to use in order to understand if our model of the
universe was correct.

The Greek philosopher Aristotle saw that during a
lunar eclipse, the shadow the Earth projected on the Moon
was always curved. From this he reasoned that the Earth
must be round (2,000 years before Columbus). Since we do
not feel the Earth move under our feet (nor are we blown

off the Earth like a rider's hat on horseback), he reasoned that we alone must be stationary. The Moon, Sun, planets, and stars must orbit the Earth in perfectly uniform circles—because the circle is the most perfect shape, and the only logical place for the center of the universe to be is at the center of these circles. This geocentric (or Earth-centered) model of the universe explained a lot, and in many ways it was a very good scientific theory. It explained what most people could see on a daily basis and tied together a variety of phenomena. It even did an excellent job of predicting what people would see in the future. Go to any planetarium, and you can see the universe circle around you on the surface of a giant celestial sphere, just as it appears in reality. It works, but that doesn't change the fact that it is wrong (although a 2014 survey by the National Science Foundation revealed that one in four Americans was not aware of that).

Even during Aristotle's day, astronomers noticed that the planets didn't seem to move across the background stars in a simple way. We can see this most easily when watching the motion of Mars from night to night. Every twenty-six months, Mars completes a loop-the-loop across the background stars: seemingly stopping, moving backward, then stopping and moving forward again. Each of the planets performs a similar type of retrograde motion at some point in its orbit. Yet Aristotle and others refused to reject a good idea for unfortunate observations: the heavens were perfect, and so were circles. Their solution was to add complicated layers of overlapping circles to each planet's orbit causing them to spiral along in their paths through the heavens.

In the second century CE, the mathematical astronomer Claudius Ptolemaeus (better known as Ptolemy) of ancient Alexandria, in Egypt, took the geocentric model and, using the ancient eclipse observations of the Babylonians, made detailed calculations for predicting the positions of the Moon, Sun, and planets within this complex geocentric system. His *Almagest*, published in 150 CE, was the definitive word on what every astronomer needed to know up to the time of Galileo. Think about that for a moment: for 1,500 years, this was the most widely used astronomy textbook in the history of the Western world. Everyone used it because it worked (mostly), and so over the centuries it became unheard of to question the correctness of Ptolemy or of Aristotle before him. Add to their infallibility the fact that the new Christian church had adopted their Earth-centered universe as proof of our special status in God's creation, and eventually, any questioning of their astronomy took on the burden of heresy. Without the freedom to question anyone's conclusions, new discoveries die, and a single astronomy text can hold sway for over a millennium.

But there were those who rejected this philosophy. They believed that no matter how elegant the hypothesis, it is the experiment that determines what is right. Almost forgotten in the West today is the work of the tenth-century Islamic mathematician and scientist Ibn al-Haytham. From

him we have one of the earliest statements of what today
we recognize as the scientific method:

> The seeker after the truth is not one who studies the
> writings of the ancients and, following his natural dis-
> position, puts his trust in them, but rather the one who
> suspects his faith in them and questions what he gathers
> from them, the one who submits to argument and demon-
> stration, and not to the sayings of a human being whose
> nature is fraught with all kinds of imperfection and de-
> ficiency. Thus the duty of the man who investigates the
> writings of scientists, if learning the truth is his goal, is to
> make himself an enemy of all that he reads, and, applying
> his mind to the core and margins of its content, attack it
> from every side. He should also suspect himself as he per-
> forms his critical examination of it, so that he may avoid
> falling into either prejudice or leniency.

Ibn al-Haytham, known in the West as Alhazen, was
born in Basra in 965 CE in what is now Iraq. At a time
when learning in Europe was stifled, with even the learning of the ancient Greeks suppressed, the acquisition of
knowledge was considered an Islamic virtue. Early in life
Alhazen turned his mind to the study of religion, but he
was dismayed by its many contradictions and the conflicts
it engendered. Believing that these disagreements lay in
human interpretation and misunderstanding, he turned his

eyes instead to mathematics. In the world of numbers, no matter what your beliefs, no matter where you were from, mathematics always worked. The fact that he thought mathematics could be used to describe God's creation was a revolutionary idea: one that would come to be called physics.

Alhazen's skills became known beyond Basra when he claimed to have calculated how to damn the Nile River that flooded each year, an event causing massive destruction and death. Unfortunately, when the powerful caliph of Egypt had Alhazen brought to him and he was shown the mighty river firsthand, Alhazen realized he was wrong. This was a difficult spot to be in with a caliph, who, though a patron of the sciences, was also a supremely dangerous man. What could he possibly do?

In the end Alhazen did the only logical thing: he pretended to go insane. The caliph had him confined to a house where he stayed for the next ten years, and he was freed only upon the caliph's death. Although his body was imprisoned, Alhazen's mind was free to work on the mathematical exploration of the world around him: specifically, the questions of sight and the nature of the bright Egyptian light pouring in through the windows of his home.

In *The Book of Optics*, written during his time of home confinement, Alhazen used his experiments to "dispel the prevailing confusion" of how we see (and what we see) by rejecting the writings of authority and instead beginning with what was actually known and observed. From that starting place he used mathematics and logical induction to develop laws of optics and vision that could actually be tested. He refuted the prevailing claim (by Ptolemy and others) that we see by our eyes sending out some "flux" that interacts with the world and then reflects back to our eyes. If the eye was going to depend on receiving information from its surroundings, why not save a step and just propose that objects emitted or reflected light, and the eye was merely the organ by which we received that light? Experimentally it makes sense, since it is easily demonstrable that the eye can be hurt by looking at incredibly bright sources, like the Sun (the perennial fear during eclipses).

During one solar eclipse, a hole in Alhazen's window shade projected the crescent image of the partially covered Sun into his darkened room. Experimenting with candles and pinholes, he discovered how to calculate the sizes and positions of the images they projected onto screens, thereby inventing the *camera obscura*. We use this technique today when we build a box with a pinhole in one end and a screen at the other and use it to safely project the image of a partial eclipse. Although modern physics students around the world still use his methods of drawing light rays, most of us in the West aren't very familiar with Alhazen—or of the role that other early Islamic scholars played in Europe's rise out of its intellectual Dark Ages. But we can see this influence every night when we look at the sky. Our bright stars are still known by their Arabic names: Aldeberan, Altair, Alcor, Alberio. Our words "altitude" and "azimuth" (which determine the position of stars in the sky) are Arabic in origin as well and are represented by Arabic numerals, including the very concept of zero. The manipulation of these new numbers gave us "algebra," which is also Arabic.

The Latin translation of Alhazen's *Optics* was read widely in the West. But while initially European scholars would credit Alhazen's discoveries in their own research, credit for developing the method that achieved those results was not granted to him. As a result, when we think of the modern scientific method and its role in discovery, it is Johannes Kepler, who lived six hundred years after Alhazen, who most often comes to mind. Kepler was well aware of Alhazen's work and was fortunate to be living in a time when brave thinkers dared to finally question the universe of Aristotle and Ptolemy (and by extension the church). Alhazen had recognized that Ptolemy's universe of crystal spheres spinning within spheres, turning with uniform circular motions, was too complex to work in reality: "The undoubted truth is that there exist for the planetary motions true and constant configurations from which no impossibilities or contradictions follow," wrote Alhazen. "They are not the same as the configurations asserted by Ptolemy; and Ptolemy neither grasped them nor did his understanding get to imagine what they truly are."

But what that truth was, even Alhazen had no idea. Six centuries later, the heliocentric model of the solar system proposed by Nicolaus Copernicus and championed by a new generation of experiment-minded scientists, like Kepler and Galileo, proved vastly simpler than Ptolemy's complex cycles upon cycles. If all the planets, including the Earth, went around the Sun, then the retrograde motion of Mars was simply an optical illusion created by the faster-moving Earth passing the slower moving Mars. But what determined the order of the planets, their spacing and speed? Kepler thought he'd discovered the solution by hypothesizing that the size and spacing between each planet's orbit was determined by the shapes of five perfect Pythagorean solids. He instantly became famous as the first person to read the celestial blueprints of God.

Yet, just like Alhazen before him, Kepler believed that the ultimate arbiter of what was true was what was seen. Sadly, when applied to Mars, what he saw was not what he had predicted. His model was exceptionally beautiful, but its predictions were unquestionably wrong. The statesman and scientist Benjamin Franklin once wrote, "In going on with these Experiments, how many pretty systems do we build, which we soon find ourselves oblig'd to destroy!" How painful it must have been for Kepler to discard what had made him so well known. Though it took him another decade of work to finally arrive at the truth, his act of intellectual honesty has become one of the defining examples of the scientific method.

What Kepler finally found were three laws of planetary motion that, although constructed specifically for the orbit of Mars, are now known to apply to all the planets around the Sun, to the Moon as it proceeds around the Earth, and to every star in orbit around any galaxy in space. They are:

1. All planets orbit the Sun in an ellipse, with the Sun located off center at one focus (a circle is just a special case of an ellipse).

The Moon orbits the Earth in an ellipse. This is why we occasionally have so-called *supermoons*, Full Moons that occur when the Moon is at its closest point to Earth. This is also why a solar eclipse two weeks before or after a supermoon is an annular eclipse, when the Moon is at its most distant point from the Earth and too small to fully cover the disk of the Sun.

2. A planet travels around its orbit at a rate that will always sweep out equal areas of space in equal amounts of time.

In other words, planets speed up when they are closest to the Sun, and slow down when they are farther away.

3. The time it takes the planet to complete one orbit around the Sun is related to the average distance between the planet and the Sun.

The constant that relates them is a product of nothing more than their mass. As a result, wherever we see two or more objects in orbit around one another, we can measure their combined mass. This single law has led to the discovery of everything from supermassive black holes at the centers of galaxies to the presence of mysterious dark matter that is the overwhelming component of the universe.

But what is the proof that these laws are correct?

Every single spacecraft that human beings have sent into space has worked because of precisely calculated orbits obeying every one of Kepler's laws. The proof of these laws is evident in every photo returned from distant Pluto to our nearby Moon. Were it not for Kepler, there would not currently be three rovers on the Red Planet that Kepler studied most intently, each carrying its own tiny sundial to calibrate its cameras. And just like the ancient clock back in Venice, each MarsDial bears a small heliocentric solar system on its face, with a representation of the Sun at its center, surrounded by the elliptical orbits of Earth and Mars.

On August 19, 2013, NASA controllers had one of those rovers turn its cameras toward the Sun. Although they had performed this maneuver many times before, on this day, at precisely the moment predicted by Kepler's laws, a tiny notch in the Sun appeared. Just a small one at first, but as the rover continued to snap images, an irregular blob began to appear against the glare of the Sun's disk. Eventually, the strange silhouette stood out perfectly against the surrounding Sun: the silhouette of Phobos, a moon of Mars. After 5,000 years of marveling at solar eclipses on Earth, of deducing their cycles, predicting their appearances, and revising our hypotheses for why they occur, at that instant our robotic emissaries stood in the shadow of a Martian moon and beheld an eclipse on an alien world.

On Thursday, February 29, 1504, when I was in the Indies on the island of Jamaica in the port which is called Santa Gloria, . . . there was an eclipse of the Moon, and because the beginning was before the Sun set, I was able to note only the end, when the Moon had just returned to its brightness, and this was most surely after two and one-half hours of the night had passed by, five sandglasses most certainly.

—CHRISTOPHER COLUMBUS, *LIBRO DE LAS PROFECIAS*

CHAPTER 3

Shadows Across a Sea of Stars

As a kid visiting the Oregon coast I often wondered, "How wide is the ocean, and what is there beyond the horizon?" As I grew older and turned my sights to the night sky, I wondered something very similar: "How far away are the stars, and are there other planets there?" Even though very few of us have ever circumnavigated the globe, and no human being has ever ventured into space beyond the Moon, we do know some of the answers to these questions. Immensity isn't immeasurable. While these vast numbers may make little sense in our daily lives, we at least know they are known.

Consider what it must have been like to live in a world where this was not true: where the sense of immeasurability, the certainty of the unfathomable, was commonplace, and the thought that the world could be known was a novel idea. The philosopher Anaxagoras was born in about 500 BCE in the eastern Mediterranean on what is now the coast of Turkey. It was a time when philosophy had only recently turned its attention to the natural world. Less than a hundred years before, Thales of Miletus supposedly predicted the solar eclipse that ended a war, thus implying that our world was predictable and events were not just the random whims of the gods.

In such a world of physical phenomena, Anaxagoras was the first, as far as we know, to understand that eclipses occur when one heavenly body blocks the light from another. This rejection of gods and dragons as the causes of eclipses was a revolutionary thought by itself, but Anaxagoras took it further: If solar eclipses happened only because the Earth had moved into the shadow of the Moon, he reasoned, then the size of the shadow must tell us something about the size of the Moon. Additionally, since the Moon covered the Sun, the Sun must be farther away. Yet to appear nearly the same size, the Sun must be larger than the Moon. Herein lies the power of scientific thought: measure the extent of the shadow sweeping across the Earth, and you know the Moon must be at least as big as the shadow, and the Sun larger still. Mysticism provided no such opportunity: if eclipses occur when a demon devours the Sun, there is no reason to believe that any measurement we make here on Earth should reveal the demon's size.

On February 17, 478 BCE, the shadow of an annular eclipse spread across the Mediterranean Sea and crossed the Greek islands and peninsula of the Peloponnese, creating a "ring of fire" in the sky that was visible for almost six minutes. Anaxagoras, living in Athens, would have been living along the midline of annularity and surely would have seen the sight, but he could not, all by himself, in only six minutes, measure the size of the shadow across the countryside. And yet in a stroke of genius, he found the answer to his question: he simply went down to the seashore and asked arriving sailors what they had seen. At that time, Athens was the center of trade for ships from all over the eastern Mediterranean. If sailors at sea had seen a ring of fire in the sky, they would remember where they had been when they had seen it. The locations of all those who did and did not see the spectacle revealed the extent of the shadow across the surrounding seascape. Without going farther than the local seaport, Anaxagoras measured the Moon.

While we do not have Anaxagoras's own words as to what he concluded, we do have the writings of those who came after. Five hundred years later, the Roman historian Plutarch wrote, "Anaxagoras [says that the Moon] is as large as the Peloponnesus." Hippolytus of Rome, a third-century father of the Christian church, wrote in his *Refutation of All Heresies* that, according to Anaxagoras, "the sun exceeds the Peloponnesus in size." The story of Anaxagoras standing on the beach measuring the size of the Moon is the story of astronomy. We are a species confined to our own world (or at best, our own solar system). Yet from this one vantage spot we have had to survey the universe on whose shores we stand. To do so we have had to study eclipses, *transits* (when small things move in front of big things), and *occultations* (when big things move in front of small things).

Astronomy is made possible, in part, by the shadows that span the stars.

Standing on the celestial seashore, let's pace out our universe, starting from the world we see by day to the stars we see at night. At each step we will learn where we are and how far we've come. What is the simplest way to measure distance? We can walk. We measure distance in feet (at least in the United States), and it's no accident that a foot is about the size of our own feet. How far can a person measure precisely by walking? In the Mediterranean of the third century BCE, *bematists* were men who could walk at a precise and constant pace, and they were paid to do so. You could hire such a man to accurately measure long distances across the landscape. Bematists were used along the Egyptian Nile, which during its annual floods erased the features marking the boundaries between fields. Bematists were particularly suited to pace off the long, flat, featureless landscape along the Nile south from Alexandria to Syene, which they found to be 5,000 *stadia* apart (around 520 modern miles, depending on the exact definition of *stadia* used). We know this distance because sometime in around 240 BCE, Eratosthenes of Cyrene, the chief librarian of Alexandria, used it to find the size of the world.

Eratosthenes had heard that on the summer solstice, the noonday Sun would shine straight down a well in Syene and cast no shadow. He knew no such thing happened in Alexandria on that or any other day, so one of two possibilities must be true: either the Earth was flat and the Sun was very close (much like a cloud that hangs over one town appears far to the south as seen from another), or the Sun was far away and the Earth was round. The answer could be found by looking at the Moon during a lunar eclipse. Aristotle had already noted a century before Eratosthenes's experiment that during every lunar eclipse the Earth's shadow looked like a circle. No matter where the eclipse occurred in the sky, the shadow the Earth cast never changed. The only figure that looked the same from any direction was a sphere.

Since the Earth's shadow already confirmed the Earth was round, the only explanation for the different lengths of the Sun's shadows in Syene and Alexandria was the curvature of the Earth. From the difference in shadows and the distance between them, the circumference of the Earth was revealed.

Here's how the principle works: Imagine that one day in the Hawaiian Islands, you notice your own shadow directly at your feet, while flag poles and Tiki totems cast no shadows at all. You look overhead and there is the Sun at the zenith, the highest point in the sky. Hawaiians call it *Lahaina Noon* after the name of a town on the island of Maui where this happens twice each year. You immediately call your friend in Puerto Rico, who is unimpressed. Right at that instant, she is watching a spectacular sunset, with the Sun touching the waters of the Caribbean on the horizon. At that moment, the two of you see the Sun 90 degrees apart in the sky, that is, exactly one-quarter of a circle apart ($90°/360° = ¼$). You must therefore be a quarter of a circle, a quarter of the Earth's circumference, away from one another. Measure the distance between you, multiply by four, and you know the full distance around the Earth.

This is precisely what Eratosthenes did. At the moment that the Sun was directly overhead and shadows disappeared in Syene, he measured the lengths of shadows in Alexandria and concluded that there was a change in the Sun's position of 7.2 degrees. This difference meant that the two cities were 1/50th (7.2°/360°) of the way around the Earth from one another. Since the distance on foot between them was 5,000 stadia, the entire Earth, Eratosthenes reasoned, must be 250,000 stadia around. Depending on the precise length of a stadia, his value for the Earth's circumference may have been off by as little as 2 percent from the actual value we now know. More important than this precision, however, is the very idea that it could be done.

For Eratosthenes's method to work, it was vital that two distant observers be looking at the same thing (the Sun, in this case) at exactly the same time and see it forming different shadows in different places. But how can we be sure that any two people on Earth are looking at the same thing in the sky at exactly the same moment? Simple: use a lunar eclipse.

As early as 150 BCE, the astronomer Hipparchus suggested using lunar eclipses to determine longitude (locations east and west) around the Mediterranean. To do this accurately, however, requires paying attention to time. When the Moon moves into the shadow of the Earth and is eclipsed, everyone on the night-side of Earth sees it happen at the same instant. However, the local time at which each person sees the eclipse occur depends on the person's location east–west across the nighttime Earth. If we know the difference in local time for an event everyone experiences at once, we can tell how far east and west they must be from one another. Since it takes the Earth twenty-four hours to turn 360 degrees, the Earth must turn 15 degrees per hour. For every difference of one hour in local time, two observers must therefore be separated by 15 degrees around the Earth. If I see the lunar eclipse begin at 10:00 in the evening, while my friend sees it start at 1:00 in the morning, the difference in local time is three hours, and we must be 45 degrees around the world from one another.

Once Eratosthenes measured how many stadia there were per degree, then a difference in local time between two observers became a difference in stadia around the planet. For this reason, it would be useful to have a tool that could calculate the expected dates and times of eclipses in a known location (against which to compare the local time at which the eclipses were seen while traveling). Just such an eclipse calculator was found in 1901 in the remains of a shipwreck off the Greek island of Antikythera. What is now called the Antikythera Device is an intricate mechanism of bronze gears that have been shown to calculate various astronomical quantities, including the position of the Sun against the zodiac (and thus the date), the phase of the Moon, and the date and time of solar and lunar eclipses.

To calculate eclipses, its dials appear to have tracked the Saros and Exeligmos cycles. From matching the intervals of known eclipses in the ancient world with the intervals predicted by the markings on the dials, scholars have dated the device to sometime around the third century BCE. That's 1,500 years before metal gears would once more be used in Europe to make clocks for telling time and modeling the heavens.

No one knows if there were other copies of the Antikythera Device traveling the Mediterranean onboard other ships, but we do know that in the centuries that followed, navigators set sail with books and scrolls filled with detailed astronomical tables filled with the positions of Sun, Moon, and stars and the times of their eclipses. Christopher Columbus was just such a navigator.

Thanks to Washington Irving's 1828 biography of Columbus, most Americans learn that it was Columbus alone who believed the world was round and that the East could be reached by sailing west. In truth, it wasn't Columbus's beliefs about the shape of the Earth, but rather about the size of the Earth, that set him apart from his contemporaries.

By a combination of wishful thinking, selective observations, and a confusion of Roman for Arabic miles, Columbus believed the Earth was a quarter smaller than Eratosthenes had found. He also thought the distance from Europe to Asia was so great—greater than anyone else believed—that the open sea between the Canary Islands (off the coast of Africa) and Japan in the East was no more than 2,400 nautical miles (only one-sixth of the way around the world). The actual distance along the route he had planned is closer to 10,600 nautical miles, over halfway around and pretty close to what the best minds of the day had calculated.

It's no wonder everyone else thought he was mad. By all rights, Columbus should have been sailing out into an ocean spanning over half the planet; it was only by luck that he found the Caribbean approximately where he expected to find the eastern reaches of Asia. It must have been so confusing when nothing he found was as he expected.

When, on his fourth voyage, Columbus used a lunar eclipse to scare the local populace, the difference in time between his hourglass and what his almanac predicted for observers back in Europe, should have revealed the mistake of where he actually was. Columbus thought he was somewhere off the coast of China, 107 degrees of longitude from Spain across the sea. Since the world turns 15 degrees per hour, there should be about 7 hours and 15 minutes of difference between his local time and clocks back in Spain. In reality, Columbus was only about 71 degrees west of Spain, and so the difference in time would have been about 4 hours and 44 minutes.

Through a lucky combination of error in reading his almanac and the difficulties of keeping accurate local time using sand in an hourglass, the eclipsed Moon wound up rising over the eastern ocean almost exactly when he expected. It was a complete and utter coincidence. Had he really been 7 hours west of Spain, Columbus would have discovered Arizona. Columbus would go to his grave never knowing how lost he actually had been. Not until Sir Walter Raleigh observed a lunar eclipse from Roanoke Island off the coast of Virginia in 1584 was the width of the Atlantic Ocean actually known.

Over the next three hundred years, explorers continued to map North America using the sky. President Thomas Jefferson had captains Meriwether Lewis and William Clark trained in astronomy to find their way across the continent. On the night of January 14, 1805, in the village of the Mandan tribe in modern North Dakota, Lewis wrote in his journal: "Observed an Eclips [*sic*] of the Moon. I had no other glass to assist me in this observation but a small refracting telescope belonging to my sextant, which however was of considerable service, as it enabled me to define the edge of the moon's immage [*sic*] with much more precision than I could have done with the natural eye." Using his astronomical almanacs, Lewis calculated his correct longitude to within 85 miles.

Solar eclipses are even better tools for finding one's location (owing to the dramatically instant onset of totality). The logs of eighteenth- and nineteenth-century solar eclipse expeditions are full of detailed and rather tedious tables of times and positions on the sky of eclipses against which the towns, rivers, and mountains of continent-wide countries and globe-spanning empires were measured. In 1869, Major John Wesley Powell and the men under his command became the first Europeans (and possibly first people ever) to navigate the Colorado River by boat and to map the last remaining blank spots in America. Powell's almanac predicted a solar eclipse on their journey, but the day of the eclipse found the one-armed Civil War veteran at the bottom of the 2,000-foot-deep Grand Canyon, out of view of the Sun. George Young Bradley, a boatman and geologist on the expedition, wrote in his private journal that "Major & brother have climbed the mountain to observe the eclipse but think it almost or quite a total failure for it has rained almost or quite all the P.M. We could see the sun from camp when it was about half covered but it clouded immediately and before the cloud passed it was behind the bluffs. Major has not come in. Cannot tell whether he saw it or not. If he did we shall have our Longitude."

Due to the typical summer monsoon clouds of the canyon country, it didn't clear, and so his exact position would remain a mystery. If only the Earth had more moons, making for more frequent eclipses. Sadly, the Earth doesn't; but Jupiter does.

In January 1610, Galileo discovered four large moons around Jupiter, each one about the same size as our own Moon. The closest of these Galilean satellites, Io, orbits Jupiter once every 1.8 days, and each time around it is eclipsed, or, more accurately, occulted, by Jupiter's disk. Galileo himself realized that these occultations happened with such regularity that they could be used to find longitude on Earth. Construct a table of eclipse times as visible from

Paris, for instance, and Jupiter turns into a tiny clock in the sky. No matter where you may be, observe a Jovian eclipse through a telescope, and you immediately know the time back in Paris to compare with your own.

When King Louis XIV sought to make France the world leader of science in the late 1600s, he employed astronomers to use Jupiter's moons to make the world's most precise map of France. His astronomers traveled all over the country with their telescopes, recording their Jovian eclipses and calculating their distance from Paris. It was the most accurate map ever produced up to that time, and it revealed that many roads and distances were actually shorter than had been believed. In her history of longitude, Dava Sobel wrote that upon seeing the new map, the king "complained that he was losing more territory to his astronomers than to his enemies."

In 1793, Alexander MacKenzie and his team of ten men lugged a telescope across the breadth of what would become Canada (a decade before Lewis and Clark would do the same in the United States). They found their location along the Pacific Coast by seeing the tiny eclipse of Jupiter's largest moon, Ganymede. Over a decade later, as Lewis and Clark were on their way back from the Pacific, President Jefferson commanded Lieutenant Zebulon Pike to determine the position of the Rocky Mountains in Colorado by means of Jupiter's moons, even though the Rockies were still under Spanish rule. Spanish troops captured Pike and his men in 1806, and before letting them go, confiscated all their astronomical observations. Pike's Peak in Colorado is named in his honor; fittingly, anyone now can drive to its top with a GPS unit triangulating the car's position from satellites in space. More than two hundred years after Pike's expedition, we still know where we are and where we are going by means of astronomy. These methods of using triangulation and eclipses are not, however, limited to distances here on Earth. The same techniques Anaxagoras used so long ago reveal that the cosmos is much larger than anyone had ever believed. To understand how big, let us turn our eyes back to the shadow our planet casts and look again at a lunar eclipse.

About two hundred years after Anaxagoras wrote that the Moon was larger than the peninsula and islands of the Peloponnese, another Greek philosopher, Aristarchus, realized that when the Moon entered the Earth's shadow, we instantly saw its size in comparison to our own. Aristarchus measured the curve of the *umbra* (the darkest part of the Earth's shadow) as it crossed the face of the Moon and concluded that between 2.5 and 3 Moons would fit side by side across our shadow. When at last Eratosthenes measured the size of the Earth, the size of the Moon was suddenly known. But again, Aristarchus went further. In his *On the Sizes and Distances of the Sun and the Moon*, written in around 280 BCE, he showed that if you knew the actual size of the Moon, then you could determine how far away it must be to appear as small as it does in the sky. We do this today with high school trigonometry.

69

the distance to the Sun, Aristarchus imagined a ,ormed between the Earth, Sun, and Moon when half the Moon was lit by the Sun as viewed from Earth (and so a 90° angle existed at the corner with the Moon). Measure the angle at the Earth (the angle between Moon and Sun on the sky), and if you know the distance to the Moon, then, from trigonometry, you know the distance to the Sun. Sadly, the distances were so vast, and the angles so close to 90 degrees (and his ability to measure positions on the sky so imprecise), that there was no way his method would work in actual practice.

In the end, Aristarchus's calculations pointed to a Sun twenty times farther away than the Moon and consequently twenty times larger (since both appear the same size in the sky). The reality was different, off by an additional factor of twenty, but the fact that it revealed a Sun very much larger than the Earth was profound. If the Sun was so much larger

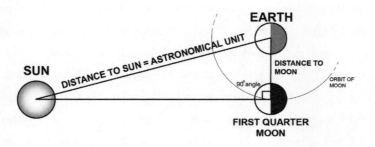

FIGURE 3.1. When the Moon is at first quarter, the Sun, Earth, and Moon form a right triangle, and trigonometry says that if you know two angles and the distance between them, the distance to the other point is uniquely known. (Image by the author)

than the Earth, then why would the Sun orbit the Earth and not the other way around? It was an early argument for those who believed in a heliocentric universe.

One proof that geocentric proponents gave for why this was impossible stated that if the Earth really did move through space, then the constellations should change size and shape with the seasons as we orbited the Sun. We would draw close to some stars and move away from others. Yet no one had ever seen this happen. But how much the stars should change their positions depended on two things: the distance to the nearest stars (whose position should change the most compared with those that were farther away); and our distance from the Sun—how far we actually moved through space as we traveled from one side of the Sun to the other (January to July, for instance).

But no one knew how far away the Sun really was. In fact, everything we know about the size of the solar system,

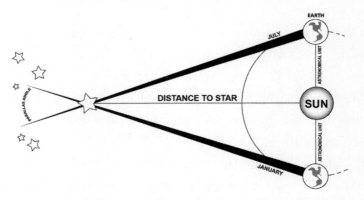

FIGURE 3.2. As the Earth orbits the Sun, a nearby star will appear to shift its position relative to more distant stars. (Image by the author)

including even the actual size of the planets, depends on knowing their distance from us. As of the nineteenth century, all of these distances were simply known in increments of the distance between Earth and Sun. For this reason, this distance was called the Astronomical Unit (AU), and its calculation was one of the great questions of astronomy in the centuries after Galileo.

The British astronomer Sir Edmond Halley was the first to suggest a means to solve this problem using a phenomenon called *parallax* on the planet Venus. On those rare occasions when Venus passes directly between the Earth and the Sun (creating a tiny eclipse, or a transit), different observers at different locations on Earth see Venus take slightly different paths across the solar disk. For observers at opposite ends of the Earth, the greater the difference in apparent paths, the closer Venus had to be. You can see this for yourself by holding your thumb a few inches in front of your face. Look at it with first one eye and then the other: your thumb shifts against the background wall depending on which eye you use and how far away you hold your thumb. The closer you hold it, the larger the shift. Our brains are hardwired to interpret the different sized shifts as differences in distance, and thus we have stereo vision that renders our world in 3-D.

Halley proposed that astronomers travel as far as possible away from each other across the globe and each record the path of Venus's transit across the Sun. Afterward, a comparison of the observations from the far-flung observers, along with simple trigonometry, would help them determine the distance to Venus. Since Kepler's third law provided the distance between all the planets in terms of Astronomical Units, once one distance was measured, all the rest would become known. Because Venus orbits the Sun once every 225 days, in principle that means that it should lap us in our race around the Sun at least once every year. But like the Moon's orbit around the Earth, Venus's orbit around the Sun is also tilted. This geometry renders those moments rare when all three bodies align. Transits of Venus come in pairs, separated by eight years. Unfortunately, over a hundred years go by between transit pairs, meaning that generations of astronomers can live and die without ever seeing one such event. Halley concluded that the next alignment would not be visible until 1761 (followed eight years later by another)—long after he would be dead. Writing in Latin so as to reach as wide an audience of scientists as possible, Halley wrote in 1716:

I would have several observations made of the same phenomenon in different parts [of the world], both for further confirmation, and lest a single observer should happen to be disappointed by the intervention of clouds from seeing what I know not if those either of the present or following age shall ever see again; and upon which, the certain and adequate solution of the noblest, and otherwise most difficult problem depends. Therefore again and again, I recommend it to the curious strenuously to apply themselves to this observation.

All over Europe, scientific societies answered his call, and as each transit opportunity approached, they launched expeditions that sailed the world for astronomy. One of the most successful transit expeditions was that of Captain James Cook, who in 1768 set sail on his first voyage into the Pacific. His destination was the island of Tahiti, where he and his crew established a small observatory at a place still known as Point Venus. He wrote: "Saturday, 3rd [June 1769]. This day proved as favourable to our purpose as we could wish. Not a Cloud was to be seen the whole day, and the Air was perfectly Clear, so that we had every advantage we could desire in observing the whole of the Passage of the planet Venus over the Sun's Disk."

On this voyage Cook would go on to explore New Zealand and the east coast of Australia, where his botanist, Sir Joseph Banks, discovered a multitude of plants and animals formerly unknown to Western science. This voyage and the two that came after made Captain Cook a hero in England and the subject of local history studies for schoolchildren, including myself, all over the Pacific Ocean and on multiple continents.

At the other end of the spectrum, and other end of the world, one of the most unsuccessful transit expeditions was surely that of the French astronomer Guillaume Joseph Hyacinthe Jean-Baptiste Le Gentil, who at the behest of the French Academy of Sciences set out for the Indian Ocean to observe the transit of 1761. Le Gentil's destination was Pondicherry, India, a site specifically suggested by Halley himself. When he arrived in the Indian Ocean, Le Gentil found that war had broken out between England and France, and the British had captured Pondicherry. Worse, storms blew his ship off course, resulting in weeks spent wandering the Indian Ocean and the Arabian Sea. On the day of the transit, he was nowhere in sight of land. He was unable to make any of his measurements accurately from the deck of a rolling ship.

Rather than returning to France in defeat, Le Gentil resolved to try again for the next transit eight years later. He set sail for Manila in the Philippine Islands, where his calculations predicted the conditions would be best for the transit of 1769. Again he was beset by misfortune. The suspicious Spanish governor there accused him of forging his letters of introduction and made it clear he was not welcome. Sailing once more for Pondicherry, now back under French control, he built his observatory and spent the time until the transit studying the local flora and fauna and Indian astronomy. Unfortunately, a storm arose on the long-awaited day of the transit, and once more Le Gentil missed everything. (The weather was perfectly clear back in Manila.) Dejected, Le Gentil packed up his samples and returned to France, but on his way home he came down with dysentery, and his ship was wrecked in a hurricane. By the time he finally reached France he had been away for eleven years, six months, and thirteen days. His heirs had declared him dead and fought over his estate. His wealth was gone, the specimens he sent from India never arrived, and his chair in the French Academy of Sciences had been bestowed on another man. According to the Canadian astronomer Helen Sawyer Hogg, who translated Le Gentil's journals, his voyage "is probably the longest astronomical expedition in history. In fact, it is quite possible that, except for interplanetary travel, there will never be astronomical expeditions to equal in duration and severity those made for that particular pair of transits."

One hundred and five years later, when the next two transits occurred, the world was better prepared than before. In 1874, Russia would launch twenty-six transit expeditions, Britain twelve, the United States eight, France and Germany six each, Italy three, and Holland one. Every country with scientific aspirations joined in the worldwide endeavor. In the United States, the composer John Philip Sousa even composed the *Transit of Venus March*.

From all these transit expeditions, astronomers calculated a distance for the AU of 92,885,000 miles, within 0.07 percent of what we now know to be the true value. This one number, acquired from observations made at far-flung points around the globe, at last revealed the extent of the solar system and each planet's size—and the numbers were huge. But let's bring these literally astronomical dimensions to a more manageable scale. Imagine we were to shrink the solar system to the size where the Sun was a grapefruit, 5 inches (10 cm) in diameter. On this scale, the Earth, no bigger than a tiny candy sprinkle 1 millimeter in size, is 16 yards (or meters) away. Jupiter, a pebble a half inch in size (1 cm), is another 60 yards away. Pluto and the Kuiper Belt, at the edge of the observable solar system, are grains of sand a third of a mile (500 meters) away from our tiny planet.

In 2006, NASA launched the New Horizons space-craft, the fastest machine ever created; even so, it took nine years to reach Pluto. Nine years to travel the third of a mile in our model solar system. Yet even light, the fastest thing in the universe, still takes almost five hours to travel that distance. The speed of light may be fast, but it isn't infinite. For centuries, philosophers debated whether light even had a speed. Some, like Alhazen in his treatise on optics, said it did, arguing that nothing could be in two places at once. Others argued that it did not. The French philosopher René Descartes wrote that if light took time to travel during a lunar eclipse, then it would take time for the Earth's shadow to fall upon the Moon, and it would take an equal amount of time for the sight of it to travel back to us. By the time we saw the eclipse occur it might already be over, and surely such a thing was impossible. Descartes was wrong, of course. Or rather, he was right; he just had the wrong moon. To truly see this effect, he needed to look for an eclipse around another planet.

In 1676, Ole Rømer's job was to measure the period of Jupiter's moon Io for the task of mapping France. Rømer, who was Danish, found that rather than having a constant rate, the time between Io's eclipses grew less as the Earth approached Jupiter and longer as it traveled away. This is so because while the Earth approaches Jupiter, the light from each little eclipse travels progressively shorter distances to reach us. The time between eclipses appears to shrink. As the Earth moves away from Jupiter, the distances increase, and the "news" of each eclipse has farther to travel. The period lengthens. The maximum difference in distance is the diameter of the Earth's orbit (twice an AU). From the maximum difference of 22 minutes in Io's orbital period, Rømer found that light requires roughly 11 minutes per AU. This is terribly fast from our terrestrial experience, but not instantaneous.

Our measurements have improved for both the speed of light and the distance to the Sun. We now know the Earth lies 8.3 light-minutes from the Sun. We see the Sun as it was 8.3 minutes ago, and we see Pluto as it looked almost 5 hours ago. But how far is it to the stars? Once the length of the AU was known, astronomers could use the parallax of nearby stars to reveal their distance. The closer a star, the more its position moves back and forth against the distant stars each year (remember the motion of your thumb when looking with one eye then the other). Because even the closest stars are so distant, the first parallax motion was not observed until 1838, long after anyone still needed proof that Copernicus had been correct that the Earth really moved around the Sun. The star was 61 Cygni, and its parallax shift was only 1/12,000th of a degree, near the limit of what even telescopes today can measure without being sent into space. The result places 61 Cygni almost 658,000

AU away from the Earth, meaning the light we see from it tonight took 10.3 years to travel here. On the scale of our model solar system, 10.3 "light-years" is the distance between Los Angeles and Paris across the surface of the globe. For comparison, the farthest human beings have ever traveled into space is the orbit of the Moon, no bigger in our model than the width of a thumbnail.

For more than a century now, astronomers have used (and built upon) this technique to erect a distance ladder out into the universe, where each step is based upon the one before. From the distance between cities in Egypt, we measure the circumference of the Earth and the locations of explorers around the globe. Their positions reveal the distance to Venus and the Sun, from which we calculate the vast distances between stars. From these distances we learn that we live in a vast spiral disk more than 100,000 light-years across containing more than 200 billion stars. In all that vastness, are there other worlds around other suns? Less than a quarter of a century ago, anyone asking how many planets there were would have been told nine (including Pluto). The number of solar systems was one: ours. Today, we know there are thousands of planets around other stars with hundreds of multi-planet solar systems. New ones are being discovered so rapidly that before I could write an exact number, it almost certainly would be out of date.

The technique that has discovered the most planets depends upon detecting their silhouettes as they pass in front of their Suns. In essence, we look for their shadows during momentary eclipses (or, more precisely, transits). The first transiting *exoplanet* (the term used for planets orbiting other stars) was discovered in 1999 around the star HD209458. For three hours every 3.5 days, it blocks enough light from its star that astronomers on Earth can tell it is 2.5 times larger than Jupiter in size.

In 2009, NASA launched the Kepler Mission, a space telescope designed specifically to look for transits around stars in a tiny patch of the sky toward the constellation of Cygnus. Kepler has found thousands of planets with thousands more waiting to be confirmed. While many of these are the size of Jupiter and Saturn, astronomers poring through the vast backlog of data are finding increasingly smaller planets. In 2015, one was found that was dubbed "Earth 2.0," a planet only 60 percent larger than the Earth in an orbit about the same as ours around a sun only slightly brighter than our own. The successor to Kepler is the Transiting Exoplanet Survey Satellite (TESS), set to launch in 2017, the same year as the Great American Solar Eclipse. Unlike Kepler, it will search the entire sky for transiting planets. When TESS finds candidates, NASA will look for ground-based observatories to conduct follow-up observations. The technology needed to do so is now available to small universities and even amateur astronomers. My students and I, using a telescope no bigger than the first one my father bought me when I was a boy, have watched a planet three times the mass of Jupiter orbiting its star every three nights, placing us briefly in a shadow more than three hundred light-years long.

The question now is whether any of these planets are suitable for life. One benefit to transits is that at the moment a planet passes in front of its star, some tiny portion of the starlight we see has filtered through the planet's atmosphere. Atoms in the planet's atmosphere absorb a tiny amount of this starlight, with each element absorbing a different combination of colors. The starlight we receive during a transit therefore bears the chemical fingerprints of the gases in the planet's sky. We are no longer simply discovering other worlds; we are learning the compositions of their atmospheres. We are, in effect, sniffing alien air.

Astronomers applied this technique to our own world. Observing our Moon during a total lunar eclipse, they were able to detect the chemical signature of oxygen, ozone, and water vapor in sunlight that had filtered through our own atmosphere before falling upon the lunar surface. Together, these molecules mark our planet as an abode for life. It's only a matter of time before we detect the same combination from planets around other stars.

That is where our steps through the universe have led: from the shores of the Peloponnese to the sands of the Nile, from the islands of the Caribbean to the village of the Mandan, and from Tahiti to the Moon, the Sun, and the inky depths of the sea of space beyond.

Someday, maybe farther in the future than Anaxagoras is in our past, the first ships will set sail for the stars. When they do, the stars that are their destinations will be the ones with worlds discovered during this generation. And they will have been discovered the way we have discovered the universe around us, by following the light and shadows of distant worlds.

By this time the light had visibly diminished: objects appeared as if lighted by the moon. The decisive moment drew near, and we waited for it with great anxiety. This, however, did not affect our intellectual powers: they were rather over-excited, and this feeling was amply justified by the grandeur of the phenomena nature had prepared for us, and by the knowledge that the fruits of our great preparations and a long voyage depended entirely upon the observation of some moments' duration.

—PIERRE JULES JANSSEN,
ON 1869 ECLIPSE EXPEDITION TO INDIA

CHAPTER 4

As Below, So Above

The Moon and Sun have not always been *places*. For the majority of the past two millennia, the prevailing view was that stars and planets were mere points of light and the Moon and Sun ethereal spheres. These spheres were eternally bright and moved each day in perfect circles above the corruptible Earth, where everything falls and decays. In matter and motion, there was nothing similar between Heaven and Earth. In the world before Copernicus, the Earth was the only *world*. This separation of Heaven from Earth and gods from man is attributable to Aristotle more than 2,300 years ago. It's an idea that makes a certain amount of sense: We see rocks tumble, yet the Sun rises. Wood burns like the Sun, yet the Sun seemingly burns forever. In fact, according to Greek mythology, we humans only have fire on Earth to fuel our civilization by the gift of Prometheus, who stole it from the heavens.

From myth to modern astrophysics, the Sun is the unifying thread in the story of our position and origin in the

Cosmos. Understanding its fire has revealed that the heavens, the Earth, and we ourselves are all subject to the same forces and composed of the same substances. This story, that we are the heavens made manifest, is one we know from eclipses.

It is a story that begins with the Moon. Though the rest of the sky is full of pinpoint stars and a (blindingly) featureless Sun, the Moon has obvious features. All over the world, people have seen these dark markings, identifying them as depicting a man, a woman, or even a rabbit. Yet up until only the past four hundred years, had you asked a learned philosopher about the nature of the Moon, you would have been told that it was a perfectly smooth, unblemished sphere. Philosophers posited that the Moon was smooth but composed of a translucent crystal with pockets of different densities. Others believed that these markings were simply the mirror-like reflection of the imperfect Earth. Some argued that the Moon wasn't solid at all, but rather a cloud of spherical vapor through which light more or less easily passed.

The unifying theme was that the Moon, as a celestial orb, must remain a perfect sphere that shared no common feature with a physical place like the corruptible Earth. Unfortunately, what we could see with our eyes could only tell us so much. That changed in November 1609 when Galileo Galilei pointed a telescope at the sky and beheld the Moon with a magnification beyond that of our own eyes. What he saw was a complex array of light and dark features even smaller and more intricate than the face we see in the Moon each month. But what could they be?* From the time of Leonardo da Vinci, Italian artists had endeavored to create realistic representations of the natural world in drawing and painting using geometry and perspective. Galileo, raised in the heart of Renaissance Florence, was intimately aware of these techniques, including that of *chiaroscuro*, the contrasting interplay of light and shadow across solid surfaces. Looking through his eyepiece, he instantly recognized the three-dimensional reality that had been hidden there for so long: "[I] have been led to the conclusion that we certainly see the surface of the Moon to be not smooth, even, and perfectly spherical as the great crowd of philosophers have believed about this and other heavenly bodies, but, on the contrary, to be uneven, rough, and crowded with depressions and bulges . . . like the face of the Earth itself which is marked here and there by chains of mountains and depths of valleys." Because Galileo does not say what curiosity made him first point his telescope at the Moon, it is impossible to say what he expected to see. As an astronomer who has had the pleasure of using a new telescope with capabilities never before available, I can say that this is the most exciting moment in any scientist's life. To look with

---

* The British mathematician Thomas Harriot actually pointed a telescope at the Moon four months before Galileo. But he made no more than a rough sketch of what he saw, noting the date and time and nothing more. What he thought about the light and dark regions that he saw he wrote in no journal; nor did he share it with any colleague. As a result, his name remains little more than a footnote in the history of astronomy.

new eyes at worlds never before seen, where every sight is a source of surprise, is science at its most thrilling.

Galileo went further: if the Moon had mountains, then he should be able to measure their height. To understand how this is possible, imagine standing in a meadow before dawn, with mountains behind you to the west. Long before you see the Sun rise, the mountain peaks behind you are lit; the first peaks to be touched are the tallest. Galileo saw the same phenomenon on the Moon. He measured how far into the lunar night a mountain could be and still have its peak lit by the Sun. The farther into night the sun-lit peaks appeared, the taller the mountain must be. His eyepiece revealed that they were comparable in height to our own mountains. Eighty years later, when Isaac Newton published his mathematical law of gravity, he proved that the exact same force that caused stones to fall also kept the Moon in orbit around the Earth, as predicted by Kepler's three laws of orbital motion. The same physical force was therefore at work shaping the landscape both here and in the sky. The Moon was now a place where any scientist or artist could imagine standing and watching the sunrise.

Could we make the same measurements for the Sun? Using a telescope to project an image of the Sun upon a piece of paper, Galileo saw that it had spots. The Sun, like the Earth and the Moon, was blemished.

Galileo and his contemporaries were not the first people ever to see sunspots. On occasion, sunspots can grow to over ten times the size of the Earth (as large as the planet Jupiter).

When this happens, they are large enough to be seen without a telescope, especially through thick haze or when the Sun is low on the horizon.* Chinese records going back to 165 BCE tell of sunspots viewed by the naked eye. The fourteenth-century Indian *Kashi Khanda* text describes a Sun whose face was covered by dark snakes—almost certainly sunspots—a sight that would have been visible as the morning Sun rose through the mist of the River Ganges.

Due to the blinding light of the Sun's *photosphere* (meaning, literally, its surface of light), sunspots are the only feature of the Sun that we can ever see under normal circumstances. But solar eclipses aren't normal circumstances. As when holding up a hand to block the glare of a street lamp at night, suddenly we see more for having less light to hide what is there. Solar eclipses have been the greatest tool for understanding the geography—or perhaps, the heliography—of the Sun. Before the nineteenth century, eclipses were interesting primarily for what their occurrence and duration revealed about navigation and timekeeping on Earth. In the early 1800s, the English astronomer Francis Baily was more interested in what eclipses revealed about the Sun and Moon themselves. While most astronomers merely observed eclipses that happened to pass over their homes (making drawings later to record what they could

---

* In 2003, a dozen major forest fires raged around Southern California. I was living under one of the smoke plumes and still remember looking up through the ash-darkened sky to see a blood red Sun with two black sunspots for eyes looking back down at me. It was the eeriest thing I've ever seen.

remember), Baily was willing to go wherever necessary to
see what he could discover and record his findings on the
spot as they were happening.

Born in 1774, Baily was the son of an English banker.
He was a studious young man and interested in science,
but as soon as his mercantile apprenticeship was over, he
left London for adventure in the United States. His letters
from this time read like a Robert Louis Stevenson novel: a
harrowing shipwreck on the high seas, disgust at the slave
trade in the West Indies, boating and canoeing down the
Ohio and Mississippi Rivers to New Orleans, and then
walking back to New York overland through 2,000 miles of
Indian-filled wilderness. For a while he thought of staying
in America (there was a rumored romance, according to
his friends back home), but eventually he returned to England. There he attempted to continue his life of travel by
seeking service with the East India Company in Turkey and
the Africa Association to explore the far-off Niger River.
In the end it came to nothing, and so finally he accepted a
position in London as a stockbroker.

Although Baily excelled and wrote revolutionary pa-
pers on the mathematics of interest and annuities—the
first to do so using algebraic equations and symbols—his
interests eventually turned back to science. In 1811, he calculated the date of the first recorded total solar eclipse in
the Western world: that of Thales of Miletus (which he
concluded occurred on September 30, 610 BCE). Twenty-
five years later, in 1836, Baily was president of the Royal

Astronomical Society, and his calculations revealed that the Moon's shadow would sweep across Europe on July 18, 1842. It would be the first total solar eclipse to cross the continent in 118 years, a sight virtually no living European could claim to have yet seen.

Baily had already seen an annular eclipse. In 1836, he had traveled to Scotland with his telescope to observe the event and in the process reported an unusual sight. At the precise moment that the Moon moved fully in front of the Sun, when the thin crescent of light wrapped itself around the Moon to become a ring, a complicated pattern of bright "beads" came into view where the two "horns" of the crescent were about to touch. Baily was not, however, the first to see these "beads." The Reverend Samuel Williams of Harvard University had seen something similar in 1780 during a total eclipse of the Sun on the coast of Maine. He had traveled there to set up telescopes and chronometers to measure the time and duration of totality. Unfortunately, whether through errors in his maps or his math, Williams's expedition just barely missed totality's path by a matter of miles. He got close enough, though, to record the appearance of multiple bright points of sunlight moving along the thin rim of the Moon where at maximum obscuration it skirted the edge of the Sun. Williams's failure to reach totality meant his observations were not widely read, but some astronomers have suggested that during his time in America, Baily might have become aware of Williams's description and so made special efforts to see these bright points of light for himself in Scotland.

The phenomena are now called "Baily's beads." They form at the first and final moments of totality when sunlight streams through mountains and valleys along the edge of the Moon—the very same ones Galileo first discovered—and they break up the Sun's light into rays that shimmer in and out of existence until the bright disk finally disappears. Without Galileo's discovery, there is no explanation for what everyone sees at the moment of totality.

It was the hope of seeing these beads again that most excited Baily about the 1842 total eclipse in Pavia, Italy. Sitting in a high apartment, his eye to his telescope, Baily's observations convey the excitement in the last final moments as the Sun disappeared behind the Moon:

> I was astounded by a tremendous burst of applause from the streets below, and at the *same moment* was electrified at the sight of one of the most brilliant and splendid phenomena that can well be imagined. For, at that instant the dark body of the moon was *suddenly* surrounded with a *corona*, or kind of bright *glory*, similar in shape and relative magnitude to that which painters draw round the heads of saints. . . .
>
> I had indeed anticipated the appearance of a luminous circle round the moon during the time of total obscurity: but I did not expect, from any of the accounts of preceding eclipses that I had read, to witness so magnificent an exhibition as that which took place.

The Spanish astronomer José Joaquín de Ferrer was the first person to call this radiance the *corona*, Spanish for "crown," when he saw it during the total solar eclipse of 1806 from the banks of the Hudson River in upstate New York. Ferrer measured the corona's extent across the sky and calculated that if there was a lunar atmosphere (which during the eclipse would be illuminated by sunlight the way steam from a cup of coffee might be illuminated by the morning sunlight on Earth), then it must extend 348 miles out into space, 50 times higher than our own. Although that seemed unlikely to him, he had no idea if it was true. Unfortunately, prior to the invention of photography, what one person beheld could be conveyed to another only through word or art. Even today it is difficult to do an eclipse justice, as no photograph captures the corona exactly as the human eye sees it. Yet as spectacular as the corona was, Baily found another sight even more remarkable: three large red "protuberances" surrounding the disk of the Moon. Although he felt certain that they were associated with the corona, whether a part of the Sun or Moon he could not say.

The interplay of expectation and surprise in what Baily saw, like what Galileo experienced before him, is one of the greatest delights in any scientific investigation.* Because of Baily's meticulous observations, astronomers from all over

---

* The most exciting statement a scientist can utter is, "Huh, that's odd."

Europe and the New World were eager to see if they, too, could discover something new about the Sun and Moon and these strange phenomena. To do so, according to Baily, was no longer possible by a single observer. Rather, the job of accurately recording all that there was to be seen during an eclipse required a team, each with his own independent piece of equipment dedicated to a particular phenomenon. Most importantly, if the Sun was to be understood, these teams of careful observers would need to travel the world to wherever new eclipses occurred.

Expeditions from England, France, and the United States would travel to every continent by commercial steamships, government gunboats, newly constructed continent-spanning railways, and even hot-air balloons. The astronomers who led them would climb high in the mountains of Peru in 1858, conduct observations in the Indian Himalayas in 1868, make beachheads on remote islands of the South Pacific in 1883, and travel across the sands of Algiers in 1900. Virtually every total solar eclipse that touched on solid land would be the subject of scientific expeditions for the next ninety years.

One British solar eclipse expedition in 1860 would mark the beginning of astronomy as most astronomers practice it today. Warren De la Rue was a pioneer in applying the newly invented technology of photography to astronomy, and he was determined to prove that the camera could capture the phenomena of totality just as well as the human eye. Through photographs, the mysterious features could finally be available for everyone to study and measure at

leisure without depending on the artistic (or literary) abil-
ities of the observer. Perhaps at last their origin would be
discovered: Were the corona and red "flames" features of
the Sun, or of the Moon?

De la Rue's expedition was a major undertaking. His primary equipment was a special solar telescope with a custom-
built cast-iron mount and forty-eight glass plates that he and
his assistants carefully cleaned and packaged in London
before departure. At their destination in Spain, the entire
apparatus would be housed in a specially designed observatory that would double as a darkroom for developing plates.
In the event the photographs were a failure (for no one had
any measure for how bright the corona and prominences
were, and therefore for how long to expose the plates), the
expedition also carried a three-inch telescope to use for
drawings. To this equipment were added boxes of developing
chemicals, distilled water, engineers' and carpenters' tools,
lanterns, oil, stove, kettle, and provisions in case the local
countryside should be deficient in food (it was not). The entire expedition equipment list could be broken down into
thirty boxes weighing a total of 4,133 pounds, transported by
the British Admiralty ship HMS *Himalaya* from Plymouth,
England, to Bilbao on the coast of Spain, and then by train
to the interior town of Rivabello. It was a long way from
Galileo and Baily and their simple pens and paintbrushes.

Fortunately, De la Rue's photographs during totality
were a complete success. From plates taken at the start and
end of totality, it was obvious that the Moon moved *across*

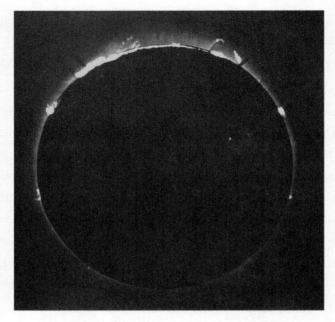

FIGURE 4.1. The first photograph of the totally
eclipsed Sun and its bright prominences, captured
by Warren De la Rue in 1860. (Image scan from the
UCLA Library Collection, courtesy Royal Astronomical Society)

the corona and prominences. The conclusion was clear:
the red "flames" and corona belonged to the Sun, and they
must dwarf the Earth in size.

As much as they were a success for solar astronomy, De
la Rue's photographs were also a success for the impartial,
mechanical eye of the camera over the artistic skill and
transitory experience of the astronomer. The Australian astronomer H. C. Russell wrote about this change with some
trepidation: "In many cases the observer must stand aside

while the sensitive photographic plate takes his place and works with the power of which he is not capable. I feel sure that in a very few years the observer will be displaced altogether." But other intellectuals were sure that even with new technology there were natural limits to what we could learn about these worlds beyond our atmosphere. In 1835, the prominent French philosopher Auguste Comte wrote in his *Course de la Philosophie Positive* (*Positive Philosophy*): "We can imagine the possibility of determining the shapes of stars, their distances, their sizes, and their movements; whereas there is no means by which we will ever be able to examine their chemical composition, [or] their mineralogical structure."

This attitude seemed eminently reasonable given the astronomical distances between us and the stars, the planets, the Sun, and even our closest neighbor, the Moon. Without ever being able to travel to the heavens, the best we could ever hope to have is their light, and once you had seen (or later photographed) all that there was to be seen, what more could possibly be known?

Two hundred years earlier, Newton had discovered that hidden within sunlight were all the colors of the rainbow. All that it took to unlock them was passing the light through a simple glass prism. In 1800, William Herschel found that if you place a thermometer just beyond the red end of the rainbow, there is an unseen "color" there that contains a tremendous amount of heat: we now call it the infrared. A year later the ultraviolet was discovered beyond

the blue. Over the next two decades, tiny gaps were discovered all across the solar spectrum where no color fell.
Because the spectrum was best viewed when light passed
through a narrow straight slit, these dark gaps were called
"lines." In 1859, just twenty-four years after Comte made his
grand pronouncement, the chemists Gustav Kirchhoff and
Robert Bunsen discovered that these lines were the fingerprints of the natural elements. Determine what fingerprint
goes with what element, and suddenly anyone can sample
the composition of the stars, no matter how far away.

We live in a world of seemingly endless variety. Yet everything we see is made of molecules that are a combination of only about one hundred different types of atoms.
Even simpler, each of those atoms is nothing more than a
combination of just three components: protons, neutrons,
and electrons. Every atom consists of a nucleus composed of
one or more positively charged protons (with some number
of neutrally charged neutrons), circled by one or more nega-
tively charged electrons. One proton circled by one electron
is all that you need to make hydrogen, the lightest element
there is. Six protons, fused together with six neutrons and
encircled by six electrons, form carbon, the primary component of our bodies. Eight protons, eight neutrons, and eight
electrons produce the most important constituent of the air
we breathe: oxygen.

We picture electrons orbiting the nucleus as if they were
tiny planets orbiting a sun. But unlike planets, electrons
can only occupy specific orbits carrying specific amounts

of energy. Any change from one orbit to another therefore requires a similarly specific change in energy. When an electron goes from a higher energy to a lower one, the excess energy is given off as light. The wavelength, or color, of light is simply a matter of its energy. Since each element has a unique set of energy levels for its electrons, each element emits or absorbs a unique spectrum of colors—its own elemental fingerprint.

Astronomers Pierre Jules Janssen and Norman Lockyer[*] independently designed instruments to study these spectral lines from the Sun during eclipses. Of particular interest were the nature and composition of the mysterious red prominences. They found that the light of the prominences, rather than being a rainbow of light like the photosphere, was instead composed of just four distinct lines: two bright red and blue lines and two weaker yellow and green. Three of those lines—red, green, and blue—had previously been discovered in the spectrum of the simplest element of all: hydrogen. August Comte had been proved wrong. Without leaving our planet, Lockyer and Janssen discovered the composition of a world out in space. Rather than some celestial ether (the dream of the ancient Aristotelians), it was nothing more than the simplest of all elements found right here on Earth. Solar prominences, it was discovered, are just great geysers of hydrogen gas erupting off the surface of the Sun.

---

[*] Lockyer was also the first person to study ancient temples and monuments, including Stonehenge, for astronomical alignments.

As for the other spectral line, when Lockyer found no element or conditions under which hydrogen emitted the remaining bright yellow line in the solar prominences, he proposed a new element named after the Sun god Helios: helium. That was in 1871. It would take another twenty-four years before this new element was discovered on Earth.

Helium, composed of two protons, two neutrons, and two electrons, is the second most abundant element in the Sun (right after hydrogen). Like hydrogen, Lockyer and other astronomers found it in the spectra of stars and gaseous clouds throughout space. The fact that the gases in our laboratory emit precisely the same spectral lines seen in the most distant galaxies—whose light has been traveling toward us since soon after the universe formed—means that the laws of physics are the same throughout space and time. As it is on Earth, so it is in the heavens.

That we see the presence of these common elements throughout space is an important clue to the life cycles of stars—and the origin of just about everything. To understand why, we must step back and first wonder why the Sun, like the stars at night, continues to shine. Is it possible that the Sun shines because it is on fire? If so, the light we see is due to the breakdown of molecular bonds. Chemists of the nineteenth century knew well what kind of energy molecular bonds released when they broke; given the rate at which the Sun shone, they found that if this were true, the Sun would run out of fuel in no more than a few thousand years. Such a short lifespan for the Sun was not a problem for a biblical origin of the Earth only 6,000 years ago. But the nineteenth century also saw expeditions other than those for eclipses. Charles Darwin's expeditions on the HMS *Beagle* in the 1830s produced his work proposing the slow evolution of species over millions of years. This theory agreed well with his calculation, presented in the *Origin of Species*, that erosion would need 300 million years to create the geologic landscape he saw back home in England. It made sense that the Earth should be older than the life that inhabited it, but it would be impossible for the Earth to be older than the Sun that sustained it. The Sun couldn't be on fire.

Did the Sun shine, instead, by the energy it released from its formation? Drop a rock from a high building, and there was no doubt that its impact released energy. An entire Sun's worth of mass falling together from a distance spanning the solar system would release enough energy to light the Sun for nearly 20 million years. Yet even that was too short compared to Darwin's geologic history. No adequate answer for why the Sun could shine for so long was known until the dawn of the twentieth century. Only then did the discovery of radioactivity and the components of the atom finally reveal a previously unknown energy source: fusion. Fuse small atomic nuclei together to form larger ones, and the energy released per mass is more than almost any other reaction.

Positively charged protons reside in the nucleus of the atom. To force them together, you need to overcome the tendency of like-charges to repel. Only incredibly high pressures and temperatures can fuse even the smallest atomic nuclei together. In 1920, the British astronomer Arthur Eddington (famous for an eclipse expedition of his own the previous year) accurately proposed that the only place these conditions were found naturally was in the hearts of the stars. Fusion was what fueled the Sun.

Our Sun is a mass of hydrogen gas a million times greater than the Earth. Gravity, the force that draws all things together, causes the Sun to collapse. But as it does, the pressure at its center grows until the hydrogen fuses to form helium, along with a tremendous amount of nuclear energy. The superheated gas swells like a hot-air balloon until it counters the gravitational collapse. The Sun, like every other star in the sky, is in a delicate balance between gravity pushing in and nuclear-driven heat pushing out. Eventually, the heat from the core, radiating out through the Sun, bubbles to the surface in enormous convection cells, like a rolling boil of water heated from beneath on a kitchen stove. The tops of these cells radiate their energy away into space as light. This is the photosphere that we see.

On Earth we are familiar with everyday things like oven coils and charcoal fires that give off light because they are hot. We call this thermal radiation. You and I are warm enough to give off infrared light, but not so warm that we glow at night with the lights off. The Sun, the stars, and the filaments inside incandescent lightbulbs are warm enough—that is why they light our world. The Sun's photosphere is at a temperature of nearly 10,000°F (5,500°C) and thus radiates light in the visible part of the spectrum.[*] The dark sunspots Galileo first saw are slightly cooler places within the photosphere. They form as different parts of the Sun rotate at different speeds (gas at the poles takes thirty days to make one trip around the Sun, while gas at the equator takes twenty-four). The Sun's magnetic field twists and knots during this differential rotation until kinks pop out of the surface like a rubber-band twisted too much. Where this happens, the magnetic fields push the flowing hot gas aside and we see down into comparatively cooler, darker gas beneath.

Hydrogen atoms stream along the magnetic field lines that loop out of these spots, emitting the bright red line of hydrogen gas that first revealed their composition during eclipses. These are the prominences that together with sunspots increase in number as the kinks in the Sun's magnetic activity ebb and flow over an eleven-year cycle. See an eclipse at the peak of the Sun's activity, when solar magnetic fields are fully knotted and the surface is covered in spots, and you have the chance to see lots of bright red prominences. Eclipses halfway between those years tend to have fewer.[†]

The source of the hydrogen we see in the prominences is a thin atmosphere of excited gas just above the

---

[*] Cooler gases just above these convection cells absorb their characteristic fingerprints of color from the light streaming by and thus are the source of the "gaps" seen in the light emitted by the Sun.

[†] The 2017 solar eclipse will be at minimum solar activity. The 2024 solar eclipse that crosses the eastern United States will be near solar maximum.

photosphere. The red light of excited hydrogen gives this layer its color, which is also sometimes visible during solar eclipses, and is why we call it the *chromosphere*: *chromo* is Greek for "color." Strangely, the chromosphere grows hotter the farther it extends from the Sun. Why it does so is still a subject of research. The leading hypothesis says that energy flows upward along the magnetic fields and is funneled into the upper atmosphere, particularly the corona above it, which can reach temperatures of millions of degrees.

Spectral lines given off by iron atoms that have had thirteen of their electrons ripped from around them are evidence of the extreme temperatures found in the corona. Until relatively recently, no lab on Earth could produce the conditions necessary for these lines. Astronomers at the turn of the last century thought they had discovered yet another new element in the Sun, which they called "coronium." Only with the new understanding of atomic physics that developed during the early 1900s was the true origin of these lines, and thus the enormous temperatures of the corona, made clear.

Think about what these temperatures mean. Every star you see in the sky is glowing because it is a few thousand degrees hot. The very hottest stars are a few tens of thousands of degrees. But when you see a total solar eclipse, that corona you witness is millions of degrees hot; it is the hottest thing the human eye will ever see in nature. Yet it comes from a place so diffuse that the light it gives off is too faint to be seen unless the rest of the Sun's light is

extinguished. This is what our Sun is: a nuclear fusion re-
actor that has been producing helium and energy (and thus
giving us life) for almost 5 billion years. It will continue to
do so for another 5 billion more. But what happens then?

The British astronomer Fred Hoyle (who earlier hypothesized the eclipse-predicting potential of Stonehenge's
Aubrey holes) was the first to propose that once a star has
run through its hydrogen fuel, the inward force of gravity
eventually forces the fusion of helium into carbon, oxygen,
and other elements in the periodic table. Since each new
element requires even greater gravitational pressures to
force them to fuse, the initial mass of each star determines
the end product after which no new fusion is possible. For
our Sun, the last element formed is carbon, after which its
core collapses into a white dwarf star, a tiny ball of carbon
atoms no larger than the Earth; the rest of its gases are
blown out into space. For the most massive stars, where elements as complex as iron are made, the end of their nuclear
production occurs in an explosion as bright as a billion stars
shining all at once. During this supernova explosion, the
atoms in the star collapse and then rebound, ripping the
star apart from the inside out. For a fraction of a second,
it becomes the universe's largest particle collider, producing the other naturally occurring elements in our periodic
tables. The star's explosion scatters all of these atoms into
space, and eventually they become incorporated into new
stars (and the planets that form around them). The spectral
lines of carbon, oxygen, and iron that we see in the Sun are

only there because other stars lived and died long before ours ever formed. They are the source of the lead in our batteries, the silver in our banks, and the uranium in our warheads. The iron in our blood was formed in the ancient hearts of stars, and with every breath we take we breathe in the oxygen those stars left for us. They are a part of us: every atom in your body, other than hydrogen, was once an atom in the heart of a star. As the astronomer Carl Sagan said, "We are star stuff."

This is a revelation both uplifting and humbling. It is the dream of the astrologers that we are intimately tied to the stars at an atomic level. At the same time it is a disquieting thought that the physicists who discovered this cosmic connection are many of them the same ones who found another, less uplifting use for nuclear processes. The physicist Hans Bethe, who discovered how helium is fused, went on to become the head of the theoretical branch of the Manhattan Project, which helped to develop the first nuclear bomb in World War II. Edward Teller, who discovered the energies produced in the nuclear fusion at work in stars, went on to do the same for the even more powerful hydrogen bombs. The fathers of stellar fusion are the fathers of the atomic and hydrogen bombs—bombs that for a brief moment unleash the conditions at the core of our Sun on the surface of our tiny planet. How appropriate then that Zeus, in his anger at Prometheus for stealing the celestial fire, sent evil into the world locked in a box. Like Pandora, in our curiosity we opened it.

But there is another use for nuclear fusion on Earth. Should we ever find a way to safely reproduce the nuclear fusion in the Sun, using the hydrogen in simple seawater, we will unleash the power of the stars using a process that leaves behind no radiation or greenhouse gases. We would be using a fuel that is found everywhere, virtually free, and practically limitless.

Eddington spoke presciently when he said, in 1920: "If, indeed, the sub-atomic energy in the stars is being freely used to maintain their great furnaces, it seems to bring a little nearer to fulfillment our dream of controlling this latent power for the well-being of the human race—or its suicide." It is important to remember, then, that although evil may have escaped from Pandora's Box, the one thing that didn't was Hope. After 5 billion years, we are the universe on Earth: the stars made sentient.

I wish to again put on record, that, during totality of the solar eclipse of 1878 at Denver, Colorado, I saw two new stars which I have reason to think were intra-Mercurial planets. . . . These are facts, and the world is challenged to disprove them.

—Lewis Swift, American amateur astronomer, 1883

CHAPTER 5

The Eclipse That
Changed the World

The first planet discovered since antiquity was found by mistake. Its discovery resulted in the greatest triumph of Newton's clockwork universe, which eventually led to its greatest failure at the hands of a German patent clerk. Out of the ashes of that failure a new paradigm would rise, not just for scientists, but for the world as we know it—and it was a solar eclipse that provided the proof.

On a clear night in 1781, William Herschel, a musician and self-taught astronomer born in Germany but raised in England, pointed a telescope at a relatively unremarkable piece of sky. In his eyepiece he saw a faint bluish ball. At first, he took it to be a new comet. But as other astronomers turned their telescopes toward the new object, its changing position revealed motion more like a planet around the Sun than a comet plunging toward it. They named the new planet Uranus, after the Greek god of the sky.

But something about Uranus wasn't quite right. No single orbital solution to Newton's law of gravity fit all of its observed positions. At the Paris Observatory, Urbain-Jean-Joseph Le Verrier was given the task of reconciling these observations. Le Verrier was a brilliant young mathematical astronomer (what today we might call a theoretical astrophysicist), and in 1846 he published a paper with a startling hypothesis: Uranus was not alone.

For almost two hundred years, Newton's law of gravity worked spectacularly well at explaining the motion of everything from a cannonball's arc to the movement of the Moon and stars. Everything about gravity was mathematically predictable. According to Le Verrier's calculations, the only way Uranus could move as it did was if it were influenced by an as yet undiscovered planet even farther from the Sun. To assert was one thing; to reveal, another. Le Verrier worked backward from effect to cause and on August 31, 1846, announced the precise celestial coordinates where astronomers should turn their telescopes to reveal the unseen planet. Less than a month later, Neptune was found precisely where Le Verrier said it would be.

Neptune's discovery was the highest triumph yet of Newton and science; it was confirmation of the contemporary assertion by philosophers that if one could but know all the "forces that set nature in motion, and all positions of all items of which nature is composed," then the future and past could be predicted with infinite precision. And such a world of mathematical precision it was. Life in the

Red prominences of hydrogen gas erupt off the Sun along with two bright "Baily's Beads," where sunlight shines through mountain valleys along the edge of the darkened disk of the Moon. Only during a total solar eclipse is the Sun's corona ever visible to the human eye on Earth. (Image copyright 1991, Fred Espenak, MrEclipse.com)

The Full Moon darkens as it passes into and out of the Earth's shadow during a total lunar eclipse. It turns orange as the only light that falls upon it is filtered through our atmosphere. (Image by the author)

A portion of the 1,000-year-old Mayan Dresden Codex. The red and black bars and dots surrounding the unique half-black, half-white figures being eaten by a serpent are numbers in the Mayan base-20 system of counting. They represent 177 and 148 (6 and 5 lunar months, respectively), the number of days between eclipse seasons. (Image courtesy SLUB Dresden, Mscr.Dresd.R.310, http://digital.slub-dresden.de/id280742827, [CC-BY-SA 4.0])

An annular eclipse of the Sun by Mars's moon Phobos as photographed from the surface of the Red Planet by the NASA rover *Curiosity* on August 20, 2013. (Image courtesy NASA / JPL-Caltech / Malin Space Science Systems / Texas A&M University)

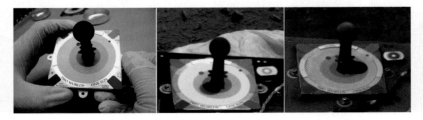

A sundial sits on each NASA rover on Mars today. The "MarsDial" (left) in the lab, (middle) on Mars soon after landing, (right) covered in Martian dust after five miles of driving. (Image courtesy NASA / JPL / Cornell University)

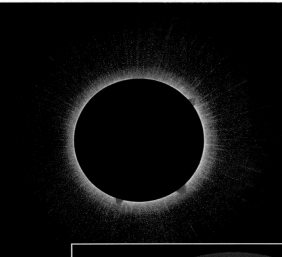

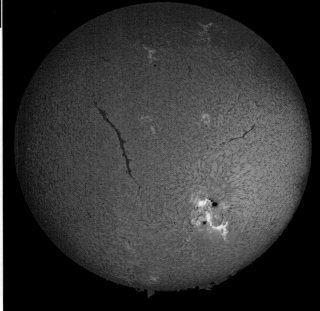

A false-color image of the Sun photographed in the red light of hydro-gen gas, revealing a group of dark sunspots toward the lower right and numerous prominences erupting off the surface along the lower limb of the Sun. The dark tendril along the center of the Sun is a prominence erupting straight toward the camera, while the bright arc between the sunspots is a rare solar flare, an explosion of exceptional power due to the sudden release of energy within the Sun's magnetic fields. (Image cour-tesy Paul B. Jones)

A petroglyph in Chaco Culture National Historical Park in northwestern New Mexico, thought to represent the total solar eclipse visible from there during the height of Chacoan culture on July 11, 1097. The curling lines may depict the looping tendrils of the corona, and possibly a solar eruption, that would have been visible during totality. (Image by the author)

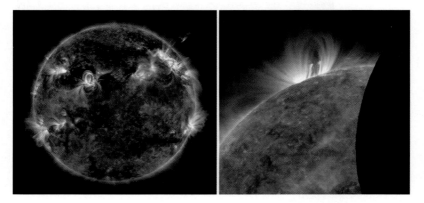

Spacecraft image of the Sun, using ultraviolet light from iron atoms at a temperature of 1 million degrees looping along magnetic field lines. Sunspots are sources of high magnetic activity and so appear to be bright. On the right, the Solar Dynamic Observatory spacecraft captures a partial lunar eclipse from space. The black silhouette of the Moon reveals the jagged edge of lunar mountains, which are dwarfed by a close-up of an active solar region larger than the entire Earth. (Image courtesy NASA/SDO and the AIA, EVE, and HMI science teams)

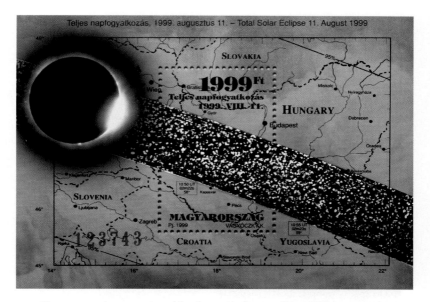

Commemorative stamp issued by the national post office of Hungary on the occasion of the August 11, 1999, total solar eclipse across Europe. Inset shows the "diamond ring" as seen at the onset of totality during the eclipse from central Hungary. (Image and stamp courtesy of the author)

The total solar eclipse of November 3, 2013, was visible to passengers off the coast of North Africa aboard the sailing ship *Star Flyer*. Both ship and shadow intersected for forty seconds as they crossed the Atlantic Ocean going in opposite directions. (Image by the author)

Four commemorative travel posters created for four different solar eclipses: annular eclipse from Chaco Culture National Historical Park, 2012 (top left); total eclipse during mid-Atlantic crossing, 2013; total eclipse from high-altitude near-supersonic aircraft above the Faroe Islands, 2015; and the "All-American" total solar eclipse of 2017. (Posters created by the author)

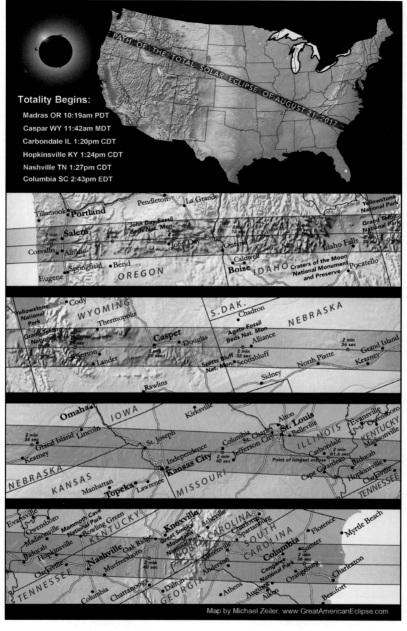

Map of the path of totality for the August 21, 2017, total solar eclipse crossing the United States. Maximum durations of totality along the midline of the path are given at four-second intervals. (Map courtesy Michael Zeiler, www.GreatAmericanEclipse.com)

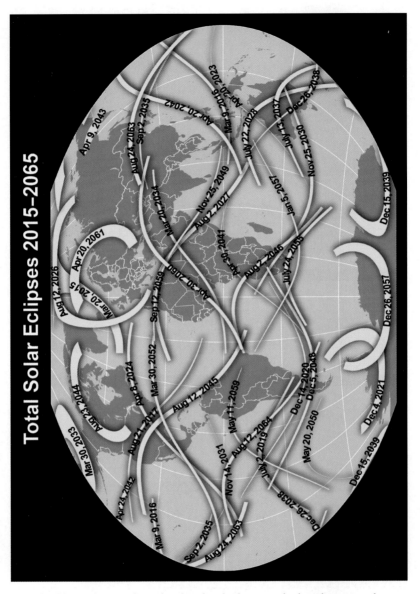

World map showing the paths of totality for future total solar eclipses over the fifty years from 2015 to 2065. See Table 7.1 for durations. (Map courtesy Michael Zeiler, www.GreatAmericanEclipse.com)

nineteenth century was accelerating rapidly, fueled by the new technology of iron and steam made possible by the physics of heat and motion. Smoking mechanical ships now crossed the seas in mere days, while locomotives raced across continents, for the first time moving people faster than any creature could walk or run. In Europe and the Americas, electricity and the telegraph made the world smaller than its physical geography and connected any two points at the speed of light (or at least the speed of a telegrapher's typing). In this world, Le Verrier's name, along with those of other French luminaries of science and technology, would be engraved in the steel girders of the tower that Gustave Eiffel would eventually erect over the city of Paris.

But Le Verrier's work did not end with Neptune; at the other end of the solar system he saw that there was something wrong with Mercury. Like the planet Venus, Mercury periodically passes in front of the solar disk. But where a transit of Venus is rare, Mercury's transits are common, happening at intervals between three and thirteen years apart. Le Verrier realized that for all of Mercury's transits, the times that had been predicted didn't *quite* match what had actually been seen. The differences were small, only a few seconds—easily explained by errors in an individual observer's clock. The problem was that nearly every observer reported this error, and all in the same direction (every transit always starting early). What's more, over the century that they'd been observed, the discrepancies had been growing. For the man who was Newton's champion, the mathematical solution was clear: our solar system must contain yet another planet, this time one hidden in the glare of the Sun. Just as Mercury's transits revealed its influence, perhaps transits of the mystery planet would reveal its existence.

Almost immediately upon announcing his conclusion, Le Verrier received word that the mystery planet had already been seen. Dr. Edmond Modeste Lescarbault, an obscure French physician living outside Paris, claimed to have seen just such a small circular spot pass across the Sun. Visiting Lescarbault incognito, Le Verrier cross-examined the doctor, "whose means of observation were certainly of the scantiest," he later reported. "His telescope was a small one of only two or three inches aperture; as a timekeeper he had only an old watch with no second-hand, so that he was obliged to use a pendulum consisting of a bullet at the end of a string for counting seconds, and to save paper he made his calculations on boards which he planed off whenever he wished to erase an old computation and make way for a new."

At the end of his visit with Lescarbault, Le Verrier announced to the world that his predicted planet had been seen. He named it Vulcan after the Greek god of fire. Using Lescarbault's time and duration of transit, Le Verrier calculated that it must circle the Sun once every nineteen days and seven hours, and that during roughly half of all solar eclipses it would appear within 8 degrees of the Sun and as bright as a first-magnitude star (one of the brightest in the sky). In addition, twice each year—around April and October—it would be visible transiting across the solar disk.

The next transit would occur on or about March 22, 1860, just two and a half months later. To find such a planet, astronomers all over the world would need to monitor the Sun constantly without any gaps in their observations, lest the planet pass across the Sun unseen. By telegraph the news went out to far-flung observers. Yet no such transit was ever seen. Even worse, an astronomer in Brazil published a paper claiming to have observed the Sun at exactly the same time as Lescarbault and to have seen no transit at all. This would be a pattern repeated over and over again for the next fifty years: for every astronomer (noted and otherwise) who chanced to see a dark dot on the Sun "exactly" as Vulcan would appear, another reported fruitless hours spent looking for any such spot. Detecting Vulcan during a total solar eclipse proved no easier.

On August 7, 1869, the *Sioux City Daily Times* reported that a party of four astronomy enthusiasts in St. Paul County, Iowa, claimed to have seen a "star" one-sixth the size of Mercury near the totally eclipsed Sun. However, the *Des Moines State Register* announced that a professional astronomer who had set up his observatory in the path of totality had "searched that region thoroughly, and found nothing that would indicate the existence of planets of that kind." The disagreement over Vulcan's existence eventually became so absurd that, according to the *New York Times*, a positive opinion of Vulcan was a dangerous matter for any young astronomer. The planet Vulcan had become "an astronomical sea-serpent." Although it might exist, the older, more established astronomers would scoff that "Professor So and So never saw it," and then they would hint, "with sneering astronomic smiles, that too much tea sometimes plays strange pranks with the imagination, and that an astronomer who cannot tell a planet from a fly that walks across his object-glass is no sort of man from whom any discoveries of moment need be expected."

All that appeared to change on July 29, 1878, when two independent astronomers claimed to have seen Vulcan during a total solar eclipse from two widely separated locations within minutes of one another. Reporting from Rawlins, Wyoming Territory, Professor James Craig Watson of the University of Michigan reported, "I had committed to memory the relative positions of the stars in the neighborhood of the sun, and I had placed the chart of the region conveniently before me for ready reference, whenever required. . . . The object which I had in the field shone with a ruddy light, and it certainly had a disc larger than the spurious disc of a star." Meanwhile Lewis Swift, the famous discoverer of no fewer than thirteen comets (including Comet Swift-Tuttle, responsible for the annual Perseid meteor shower each August), reported from totality in Denver, Colorado, that in searching for Vulcan, "about one minute after totality I observed two stars, by estimation 3° S.W. of the sun. . . . [B]y careful comparison, they appeared exactly of the same magnitude, and both as red as Mars. I looked closely for twinkling, but they were as free from it as the planet Saturn.

They both, at the time, seemed to my eye and mind, to have a small round disk, about like the planet Uranus."

Upon this news the Princeton astronomer Dr. C. A. Young wrote that "one brilliant discovery will probably develop from this occasion, and hold a conspicuous place in the annals of science. The planet Vulcan, after so long eluding the hunters, showing them from time to time only uncertain trace and signs, appears at last to have been fairly run down and captured." Given the brightness of the new planet, Young calculated Vulcan's size could be no more than four hundred miles in diameter. This was tiny enough to have escaped detection for so long, but too small to cause the change in Mercury's orbit in the way that was seen. Perhaps there were multiple Vulcans? In fact, it soon became clear from comparison of Watson's and Swift's notes that their locations didn't match—they couldn't possibly be reporting the same planet.

When no subsequent transits appeared, each man became adamant that what he had seen was real. The *New York Times* wrote, "Prof. Swift arrived in town to-night, and in an interview with a reporter stated that he had no more doubt that he discovered Vulcan than that he had been to Colorado." In time, Swift became sure that the new photographic technology being used by eclipse-chasers like Pierre Janssen would confirm his discovery. When it didn't, Swift was ready to explain that this was not surprising, as the photographic emulsions were more sensitive to blue light, and not the red color he'd seen Vulcan display. "Happily the time is not far distant when the problem can be settled," wrote Swift in 1883. "The great eclipse of 1886 will afford an admirable and comparatively easy opportunity, if rightly managed, to dispel every doubt. . . . Until then let us hold the matter in abeyance. My faith in their existence was never stronger than to-day."

But the eclipse of 1886 revealed nothing about Vulcan. As the nineteenth century gave way to the twentieth, the triumph of Newtonian physics that had led to the discovery of Neptune was now producing nothing but failure. With each eclipse that revealed no sign of the elusive planet, further excuses and refinements were required to explain away Mercury's motion. What began with a single planet had already been modified to multiple planets, which in turn became a plethora of planetoids, a belt of debris between Mercury and Venus, and finally nothing more than a simple ring of dust. Nothing ever fit. And yet still Mercury moved.

Perhaps the fault was not in our stars but rather in our physics? The astronomer Asaph Hall, discoverer of the two moons of Mars, suggested an unsettling possibility in the summer of 1894: maybe gravity didn't work the way they thought. Was it possible that Newton was wrong? For scientists, the defining characteristic of our field is the experiment. If an experiment's results do not confirm our hypothesis, it is the hypothesis that must change. This is an ideal that depends on there being a simple, crucial experiment whose result is agreed upon by all involved to be the deciding factor of which idea is correct. The reality is never this simple. It is always easier to add an extra parameter to a previously successful theory than to scrap the whole thing and start from scratch: Mercury doesn't move as expected? Simple, just keep adding ever tinier planets. But the question arises, when do you stop making excuses and look for a new hypothesis?

Occam's Razor is the supposition that the simplest explanation that fits all the data is usually the correct one. Unfortunately, there is rarely any agreement on what explanation is "simplest," and "usually correct" is not the same as "always correct." In his book *The Structure of Scientific Revolutions*, the philosopher Thomas Kuhn explains that science does not progress by a constant stream of crucial experiments, with scientists constantly reevaluating all of their assumptions and successes. Rather, scientists use the results of previous experiments to build a framework, or paradigm, upon which to hang all of their new experimental results, gradually constructing a picture of the universe. Based on this evolving picture at any given moment, scientists think of new experiments to perform and decide how to interpret their results. The majority of the time, we are simply filling in the missing pieces of a picture we have inherited from those who came before.

When experiments don't provide the results we expect (e.g., when the planet Vulcan fails to appear), we look for reasons that allow us to keep as much of the framework as we can, even if the details of the picture become more

complicated than we would like (e.g., rings of small aster-
oids especially oriented to affect the planet Mercury and
no other). Eventually, someone comes along who suggests
a completely different framework that creates an entirely
new picture—a new paradigm—by which to interpret our
previous results. Whether this new paradigm is "simpler"
than what came before, thus satisfying Occam's Razor, is
rarely agreed upon by the scientists of the time. According to Kuhn, scientists during these scientific revolutions
rarely have a rational reason for choosing one framework
over another.

For instance, Copernicus proposed a new paradigm in
which the Earth was only one planet among many in orbit
around the Sun. When Galileo's telescope revealed moons
orbiting Jupiter, and previously unknown features of our
Moon, the Sun, and the other planets of the solar system, it
refuted a tenet of the old paradigm, which had said that all
motion must center on the Earth, while confirming a central aspect of the new paradigm: that the heavenly spheres
were physical places just like the Earth. Copernicus's model
of planets orbiting the Sun in perfect circles did not, however, predict the positions of the planets as well as Ptolemy's
complex systems of celestial spheres within spheres did.
Understandably, a reasonable person could choose the old
complexity over the new simplicity when simplicity didn't
work. But new paradigms suggest new hypotheses—with
new experiments that might make no sense under the old
framework.

The parallax motion of nearby stars over the course of a year is a phenomenon that makes no sense in a universe where the Earth doesn't move. Likewise, a universe where the Earth is simply another planet requires that the Moon, planets, and Sun also exhibit features and turn on their axes just like the Earth. In time, all of these phenomena were revealed through the ever-increasing magnification of telescopes. Meanwhile, the physical laws of Kepler and Newton revealed that the same force that causes an apple to drop here on Earth makes the planets orbit the Sun in elliptical, not circular, paths, a discovery that produced even better agreement with observations than Ptolemy's crystalline spheres. Whatever reason individual scientists have for accepting one framework over another, eventually no serious scientist is left to propose new additions to the old paradigm, and the scientific revolution is complete. From that moment onward, it becomes the task of new scientists to understand the implications of new laws within the new framework.

At the dawn of the twentieth century, astronomers believed they understood the framework of Copernicus, Galileo, and Newton so completely that the well-respected astronomer A. A. Michelson could write:

The more important fundamental laws and facts of physical science have all been discovered, and these are so firmly established that the possibility of their ever being supplanted in consequence of new discoveries is exceedingly remote. Nevertheless, it has been found that there are apparent exceptions to most of these laws, and this is particularly true when the observations are pushed to a limit, i.e., whenever the circumstances of experiment are such that extreme cases can be examined. Such examination almost surely leads, not to the overthrow of the law, but to the discovery of other facts and laws whose action produces the apparent exceptions.

By this reasoning, since the famous astronomers of the nineteenth century were doing nothing more than filling in details of well-established laws, then perhaps it is no coincidence that today almost no one remembers their names. But the nineteenth century was also the greatest time of discovery for forces that had seemingly little to do with planets or gravity: electricity, magnetism, and optics. In 1865, while astronomers were busy looking for Vulcan, the Scottish physicist James Clerk Maxwell unified these phenomena into a set of four equations that revealed light to be a wave of changing electric and magnetic fields.* We tend not to think of electricity and magnetism outside of the electronics that power our daily lives, yet four of the five senses with which we perceive the world—taste, touch, sound, and smell—are just the interactions of atoms and

---

* These laws are: (1) moving charges and changing electric fields create a magnetic field; (2) changing magnetic fields create currents and electric fields; (3) electric fields are caused by charges; and (4) there is no such thing as a magnetic monopole (all magnets come with two poles).

molecules via their electric fields (while sight is the direct detection of light).

In 1895, at the age of sixteen, Albert Einstein thought very deeply about light and the ramifications of the laws that governed it. For instance, he wondered, what would a person see if he could ride a beam of light? A person on a beach sees waves crashing one after another on the shore, the water waving in and out. But a person surfing on a wave, traveling at its speed, rides a constant crest that no longer appears to "wave." Maxwell's equations provided no solution for this possibility with light. They required that for light to be "light," it must move precisely at the speed of light. In fact, Michelson and other astronomers had verified this experimentally: no matter where you looked or how you moved, light always moved past you with the same speed.

It so happened that Maxwell's equations had other problems with moving observers: charged particles in motion produced magnetic fields, while stationary ones did not. Depending on whether I am standing next to a proton or moving by one at a constant rate, I will see it produce different fields. This may not seem like a profound problem, but it puzzled Einstein and would eventually lead him to solve the mystery of Mercury—and in the process overturn how we understand the nature of time and space. The reason this is so puzzling is that the Relativity Principle, which has its origins in Galileo's efforts to prove that the Earth can move without us feeling it, states that there is no experiment that can reveal whether a person is at rest or in constant, uniform motion. This principle must be true, since even when we are at rest in a laboratory, any experiment we do there is really flying at tremendous speeds as the Earth hurtles through space. Everything is always in motion relative to something. But Maxwell's laws do not satisfy this principle.

Let's say I hold two protons while standing inside a rocket moving at a constant speed. I open my hand and, because we are at rest with respect to one another, all I see is a repulsive electric force between them (their like charges repel). They fly apart and take exactly one second to hit the surrounding wall. But a friend standing outside the rocket will see me and the two protons in motion. There is now an additional magnetic force of attraction between the protons that causes them to move apart more slowly. By her watch, the same protons now take *two* seconds to hit the wall. But by the Relativity Principle, we can't both be correct; otherwise, measuring the time it took the protons to fly apart would be a simple test to determine who was in motion and who was not. Einstein found that both observers could be correct, but only if time passed differently for people at different velocities. Moving clocks, he said, must run more slowly than those at rest; the closer a clock approached the speed of light, the slower time would run. The answer to Einstein's original question was that a person moving at the speed of light would see a light wave stop waving, because time would stop altogether.

Einstein was not the first to propose a solution where space and time were relative. Henri Poincaré, a French mathematician and philosopher, had done so in his book *La Science et l'hypothese* (*Science and Hypothesis*), which was widely read among the intelligentsia of turn-of-the-century Europe. Einstein was influenced by Poincaré (as was, apparently, the artist Pablo Picasso). Einstein, however, was the one to take these radical ideas to their natural mathematical conclusions and have the courage to declare that this was how reality worked—no matter how much it might disagree with common sense (common sense, according to Einstein, being just the collection of prejudices someone acquired by the time they were eighteen).

In 1905, Einstein published his Special Theory of Relativity specifically looking at how motion at constant velocity (a "special" case of motion) solved the problems with Maxwell's equations. The only way for the laws of electricity, magnetism, and light to work for any observer traveling at any constant speed was if the following were true:

1. Moving clocks run slowly compared with clocks at rest.
2. Moving meter-sticks are shorter compared with meter-sticks at rest.
3. Events that are simultaneous for observers at rest need not be simultaneous to observers moving at a constant speed. Simultaneity is relative.

As revolutionary as these ideas were, Einstein realized his Relativity theory was incomplete. What about *any* motion, including those where an observer was accelerating?

It was this question that brought Einstein into the realm of Newton: he realized that an extension of the Relativity Principle to acceleration required that any physical phenomenon that happened while accelerating must also happen in a sufficiently large gravitational field.

We call this the Equivalence Principle, and there is a good chance you have experienced it for yourself. If you have ever been to Disneyland, you may have seen a ride where you sit in a room and feel as if you were zooming through space in a rocket. When the view out the front "window" shows the rocket blast off, hidden hydraulics tip the room backward, letting the Earth's gravity pull you back into your seat. The pull of gravity is indistinguishable from the feeling when you accelerate, and it is the reason the ride works. The same principle is at work in movies like *2001: A Space Odyssey* or *Interstellar*, where a spinning space station produces an acceleration which the characters experience as "artificial" gravity.

Einstein spent another decade working through the mathematics of this more "general" theory of Relativity. By the time he completed his equations in 1915, he had come up with two other strange conclusions:

1. Gravity is a curvature of the fabric of a four-dimensional "spacetime" (three dimensions of space and one dimension of time).
2. Clocks close to a massive object run more slowly compared with those farther away.

Gravity was no longer a force between objects with mass, as Newton proposed, but a warping of the fabric of both space and time that reproduced all the planetary motion predicted by Newton's and Kepler's laws. If you've ever seen a coin roll into one of those giant plastic funnels at a science museum, you will have seen it sweep in and out of the "gravity well" at the bottom in a way that reproduces the elliptical orbits and changing velocities of planets around the Sun. Get very close to a massive object like the Sun, and the warpage in spacetime causes planetary orbits to curve a little more than the amount Newton's and Kepler's laws predicted. The added curvature causes a planet to overshoot the point where its orbit began, and each new orbit begins a little farther along than the one before. As a result, its orbital axis shifts a tiny amount each trip around and traces out a shape like a flower, with each orbit a petal (rather than a single constant ellipse). A person on a planet farther from the Sun, like the Earth, sees a planet closer to the Sun, like Mercury, cross the Sun earlier than Newton's and Kepler's laws would have otherwise predicted, and over time, the discrepancy grows.

Announcing his new General Theory of Relativity to the Prussian Academy of Sciences on November 18, 1915, Einstein stated, "In this work, I found an important confirmation of this radical Relativity theory; it exhibits itself namely in the secular turning of Mercury in the course of its orbital motion, as was discovered by Le Verrier. Namely, the approximately 45 arcseconds per century amount is qualitatively and quantitatively explained without the special hypotheses that he had to assume."

So much for Vulcan.[*]

"Furthermore," continued Einstein, "it shows that this theory has a stronger (doubly strong) light bending effect in consequence through the gravitational field" than what could be explained by Newton's gravity alone. In other words, just like with Mercury, light traveling close to a massive object has its path deflected by the curvature of spacetime more than could be accounted for by Newton's laws alone.

The mark of a successful scientific theory is that it ties together a wide range of physical phenomena, explaining accurately what is already seen and predicting results for experiments yet to be performed. General Relativity did exactly that. It tied together space, time, motion, light, electricity, magnetism, matter, and gravity. It explained what more than fifty years of transit and solar eclipse observations had failed to verify, and it suggested a result for a new test: the bending of starlight passing near a mass like the Sun. When could such a phenomenon be tested?

The answer was: during a total solar eclipse. The change in light would be subtle. Starlight passing through the Sun's warped gravitational field would cause the light to reach Earth from a slightly different direction than if the Sun were not

---

[*] Almost no one remembers the hypothesized planet today. In 1962, the American television writer Gene Roddenberry was drafting a story for a new science fiction TV show and his main alien crew member was labeled a "Martian." Roddenberry later changed this to something more exotic. What caused him to make Star Trek's Mr. Spock a Vulcan, he never said.

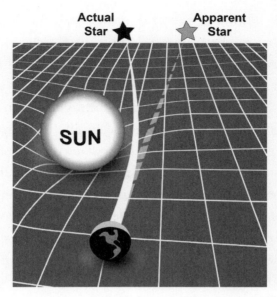

FIGURE 5.1. Starlight passing close by the Sun appears to arrive at the Earth from a different location due to the curvature of spacetime. (Image by the author)

present. Stars near the Sun on the sky would all appear slightly shifted away from its disk when compared to their known positions. A photograph of the totally eclipsed Sun would record the faint light of all the stars that were momentarily visible. Compare these positions with a photo of the same star field without the Sun (perhaps taken as little as a couple of months after the eclipse), and the change in the stars' positions should reveal any gravitational influence of the Sun.

The first attempt to measure this deflection of starlight during a solar eclipse, in 1912, failed because of bad weather.* The next opportunity occurred on August 21, 1914, from southern Russia. Unfortunately, the assassination of Archduke Ferdinand in July of that year led to the German invasion of Russia on the first of August and the start of World War I. Einstein's colleague Erwin Freundlich, who had traveled there specifically to measure the deflection of starlight, found himself instantly converted from visiting astronomer to enemy alien. He had all of his instruments confiscated and was instantly arrested by the Russians. As the United States remained officially neutral, a group of astronomers from California's Lick Observatory was allowed to remain, but they ran out of luck when the weather turned bad, and Russian officials impounded their instruments for the rest of the war.

World War I hindered the proof of Relativity in more ways. While today we may forget that Einstein was German, scientists of the day did not. A British zoologist included in his paper of 1918 a note saying, "No quotations from German authors published since August, 1914, are included. 'Hostes humani generis' [enemies of the human race]." The director of the Observatory of Turin in Italy gave lectures on the nature of "German science and Latin science" exhorting his audience to not look to German scientists "for bold initiative, for flashes of genius, for fruitful

---

* The astronomer C. D. Perrine, director of the Astronomical Observatory of Córdoba in Argentina, reported rain before, during, and after the eclipse: "We suffered a total eclipse instead of observing one."

WAR—TWO   WORLD-EMBRACING   SHADOWS—**ECLIPSE**

By a strange coincidence, at the very moment when all Europe is joined in the clash of battle the world will witness Nature's most awe-inspiring phenomenon, which in olden times —when men were "dismayed at the signs of heaven"—struck terror into all hearts. To-day (Friday) there will be a total eclipse of the sun, visible as a partial eclipse in London, where it begins at 10.59 a.m. and ends at 1.21 p.m., the greatest phase occurring at 11 minutes after noon. The portion of the earth upon which the penumbra, or partial shadow, will fall includes the area involved in the Great War. In Germany and Austria (omen faustum?) the eclipse will be nearly total.

FIGURE 5.2. Engraving from *The Graphic* magazine, 1914, illustrating the two shadows descending upon Europe—the total solar eclipse of August 21, 1914, and the near-simultaneous outbreak of war. (Image courtesy Michael Zeiler, private collection)

ideas, for results expressed in a brilliant, clear, comprehensive, and simple formula, as are all the laws of Nature. . . .

Even where hatred of Germany wasn't evident, many of the Allied scientists who heard of Einstein's theory of gravitation were sure it must be wrong. The American astronomer Heber Curtis was the leader of Lick's efforts to search for both Vulcan and any deflection of starlight. The photographic plates he acquired during the 1918 total solar eclipse in Washington State produced nothing but ambiguous results (in part because of thin clouds, but mostly thanks to cobbled together equipment replacing the pieces still impounded in Russia). Yet still he thought them worth publishing, so that, "When the Einstein theory goes into the discard, as I prophesy it will go within ten years, these negative or indecisive results will be more highly regarded than at present."

Ironically, by war's end it was a British astronomer, Sir Arthur Eddington, who was the unofficial spokesman for the German theory of Relativity. Thanks to a colleague in Holland, Eddington had the only copy of Einstein's paper to reach England during the war, and he was the primary person spreading word that the mystery of Mercury's orbit had been solved. As much as Allied scientists may have been loath to accept the broader implications of Einstein's theory, the solution to the problem of Mercury could not be ignored. As a devout Quaker, Eddington was a pacifist like Einstein during a time when pacifism was not a popular position on either side of the war. In England, those who refused to serve on the basis of conscientious objection faced possible imprisonment. According to one of Eddington's students, Subrahmanyan Chandrasekhar, it was a social disgrace to even associate with a conscientious objector; the view among the older faculty members at Cambridge University where Eddington taught was that it would bring disgrace upon the institution to even have one in their midst.

To avoid such a scandal, and to protect Eddington from the draft, "they therefore tried through the Home Office to have Eddington deferred on the grounds that he was a most distinguished scientist and that it was not in the long-range interests of Britain to have him serve in the army," Chandrasekhar later wrote. "Eddington was deferred with the express stipulation that if the war should end by 1919, then he should lead one of the two expeditions that were being planned for the purpose of verifying Einstein's prediction with regard to the gravitational deflection of light."

The eclipse on May 29, 1919, would be the best opportunity yet to test the deflection of starlight, as the Sun would be passing before the Hyades star cluster, resulting in numerous bright stars visible beside the Sun. Totality would cross the Atlantic Ocean, touching land in northern Brazil and again on the island of Principe, off the coast of western Africa. By 1919, the war had ended, but astronomers from nearly every nation that would normally mount such an expedition were out of money or simply too busy recovering from the devastation. England alone found itself prepared to set sail to test this new theory of gravity. Eddington and an assistant from Cambridge would head to Principe, while the Astronomer Royal, Frank Dyson, sent two of his assistants from the Greenwich Observatory to observe the eclipse from Brazil. Each team would take along two instruments for photographing the eclipse. Their task was to capture photos of the eclipse, as well as comparison plates of the same region of sky without the Sun present. Any change in the positions of the background stars would reveal the effects of the Sun's gravity on light.

The observations and their analysis would be difficult. The deflection Eddington was looking for was tiny: only about an arcsecond in size (an arcsecond is $1/1,800$ of the solar diameter as seen from the Earth). Any temperature, focus, or mechanical changes in the telescope could skew where the individual stars fell on a photographic plate by at least that amount. Correcting for these possibilities required taking even more comparison photos. For the team in Brazil, the eclipse would occur soon after sunrise. If they remained there an additional two months, they could photograph the same stars before dawn, with the telescope in the same position, but the Sun no longer present.

In Africa, where the eclipse would occur near noon, it would take nearly half a year for the same stars to be photographed in the same position by night. As a compromise, comparison plates were taken back in England before the expeditions set sail. This was not an ideal arrangement, but both groups were already making do with makeshift equipment pieced together from parts available in the aftermath of war. The end result was that a difficult experiment would be performed under difficult conditions where the final results, though simple in principle—a deflection of 1.75 arcseconds would confirm Einstein, a deflection of half that would support Newton—would be anything but simple in practice.

To prepare the public for the results of an experiment testing the implications of a theory that even professionals couldn't claim to understand fully, Eddington, Dyson, and their assistants implemented a public relations program aimed at both the general populace and the scientific community. It was an effort without clear precedent. For the public, they published explanatory newspaper articles about the experiments in *The Times* of London and gave public talks. For the scientific community, they wrote scholarly articles in *Observatory*, a leading professional journal, of which Eddington was the editor. These articles continued through the launch of the expedition and up to the day of the eclipse itself, with breathless reporting from the astronomers on station as the eclipse commenced.

The Times of London reported, on June 4: "Astronomer Royal (Sir Frank Dyson) informs us that he received yesterday a cablegram from Professor A. S. Eddington and Mr. Cottingham from Princes Island [*sic*], West Africa, stating that the eclipse of the sun was observed there through clouds, but they are hopeful of obtaining good results." A day later, *The Times* informed readers that "further telegrams from the British astronomers who observed the total eclipse of the sun last week report that the photographs taken at Sobral, Brazil, were quite successful, and the negatives already developed show all the stars that were expected to be recorded." Upon their return, other articles highlighted their work over the glass plates to tease out the results that would determine which great mind, Newton or Einstein, would prevail.

During this time the members of the general public weren't the only ones who needed to be properly prepared. Einstein's mathematics and the implications of his work were so complex that even Dyson could later write, "The result was contrary to my expectations, but since we obtained it I have tried to understand the Relativity business, & it is certainly very comprehensive, though elusive and difficult." Prior to making any official announcement of their results, however, Eddington first quietly presented the findings to a leading group of British mathematical physicists. Only with their favorable reception were Eddington and Dyson confident enough to announce what they'd found. They presented the findings to a joint meeting of the Royal Astronomical Society and the Royal Society two weeks later.

On November 6, 1919, in front of over a hundred members of the scientific community (as well as one reporter from *The Times*), Dyson and Eddington announced that Einstein's theory had been confirmed. Upon hearing their analysis, the president of the Royal Society declared,

If the results obtained had been only that light was affected by gravitation, it would have been of the greatest
importance. . . . But this result is not an isolated one; it
is part of a whole continent of scientific ideas affecting
the most fundamental concepts of physics. . . . The difference between the laws of gravitation of Einstein and
Newton come only in special cases. The real interest of
Einstein's theory lies not so much in his results as in the
method by which he gets them. If this theory is right, it
makes us take an entirely new view of gravitation. If it
is sustained that Einstein's reasoning holds good—and
it has survived two very severe tests in connection with
the perihelion of Mercury and the present eclipse—
then it is the result of one of the highest achievements
of human thought.

The next day the headlines blared out from *The Times*,
"REVOLUTION IN SCIENCE.—New Theory of the universe.—Newton's Ideas Overthrown." The groundwork
Eddington laid for Relativity's favorable—even ecstatic—
embrace in London caught the attention of the American
media, which had not been subjected to Eddington's public relations onslaught of the preceding year. The headline
in the next day's *New York Times* read: "LIGHTS ALL
ASKEW IN THE HEAVENS—Men of Science More
or Less Agog Over Results of Eclipse Observations—
EINSTEIN THEORY TRIUMPHS—A Book for 12 Wise

Men: No more in All the World Could Comprehend it, Said Einstein, When His Daring Publishers Accepted It."

Eddington made Einstein a genius, but the American press made Einstein a star. Due to the press coverage of the 1919 total solar eclipse, we now live in a world where Einstein's name is universally known and synonymous with brilliance, where everyone knows that black holes "suck in" light, that science-fiction starships fly through space using their "warp drive," and that "everything is relative." No scientist or philosopher predicts any longer that everything that can be known is known, or even that what is known now will always be known to be true. Even our language has changed: scientists are now careful to talk about scientific "theories" instead of "laws," even when, like the General Theory of Relativity, they have withstood nearly a century of repeated experimentation.

That repeated questioning and testing of Relativity has been vital to its success. In the years since 1919, there have been those who have claimed that given Eddington's public (and not so public) support for Relativity, he cooked the books on his "decisive" experiment, and that he threw out data that didn't match the answer he wanted. It's true the observations were not all that anyone could have wanted in a decisive experiment. Errors introduced in the stellar positions due to focus, tracking, temperature, and travel were not easily, or obviously, removed by all the comparison plates, and Eddington *did* throw out data he felt were not trustworthy.

But in a 1979 reanalysis of the plates using modern automated measurement software, the Greenwich Observatory confirmed the earlier results of 1919. More importantly, however, even after the results of 1919, the scientific community as a whole (including Eddington and Dyson) recognized the need to confirm the results during subsequent eclipses. At the next optimal eclipse in 1922, the skeptical astronomers at Lick Observatory successfully acquired results that confirmed those of Eddington. But the testing and retesting didn't end there. Such is the tenacity of science that the last professional eclipse expedition to measure the deflection of starlight was in 1973, led by a University of Texas team looking to test Einstein's theory against an even newer alternative.

Since then, radio astronomers have been able to measure the same deflection of light around the Sun using quasars—distant supermassive black holes—confirming Einstein's theory to a precision far greater than Eddington could have ever dreamed possible. Astronomers continue to subject the details of the theory to experimentation, including the detection in 2015 of gravitational waves of spacetime caused by two black holes colliding. In a fraction of a second they released ten times as much energy as the combined luminosity of every star and galaxy in the observable universe. Such waves had been predicted by Einstein in 1916, and subsequently deduced in the 1970s from the motions of two neutron stars in orbit around one another (earning two astronomers a Nobel Prize in Physics in 1993). With this new window on our universe, astronomers continue to probe the strengths and limitations of Relativity. Any weaknesses discovered in Einstein's theory in the future will raise new questions, spurring new experiments, and in time lead to even deeper understanding of the strange and magnificent universe in which we live. The story of the orbit of Mercury demonstrates the same principle: the beauty of science is that its discoveries are never the end of a story, merely the first step in a new and different one.

The perfect golden ring of light with a blazing jewel set in it was a sight that will never be forgotten by those who had the good fortune to see the eclipse yesterday morning from a place of vantage. . . . The astronomers may have further advice about the Einstein theory, or whisperings of a planet nearer the sun than Mercury, or word of a new element in the sun. But the great lesson of the eclipse to the masses of those who saw it is that one little unusual phenomenon in the skies makes us realize how closely akin we all are in this common planetary boat out on an ethereal sea that has no visible shores.

—THE NEW YORK TIMES, JANUARY 25, 1925

CHAPTER 6

Saros Siblings

The Faroe Islands are an archipelago of eighteen glacier-carved specks floating on the cold, gray waves of the North Atlantic. Their jagged mountains stretch out of the sea to catch the rain-filled clouds blown over from far-off Iceland. Everywhere I look, water tumbles down treeless slopes of volcanic rock, gathering into pools and waterfalls that pour back into the surf below. When the Sun finally does break free from the overhead gloom, the cascades come alive and burst with rainbows, while sunbeams race across the mountaintops. The British, no strangers to harsh weather, call this the Land of Maybe. Upon arrival, I have a hard time believing this is where I've traveled to see a total solar eclipse.

It won't be the first total eclipse seen here by the descendants of the Norsemen who once fished these waters. One of the oldest stories the islanders tell is of four brothers from the southernmost of the islands, Suðuroy, who may have witnessed the total eclipse of May 30, 1612:

They were brave and strong, but they were constantly quarrelling and fighting, and sometimes even threatening to take each other's lives. One day, when they were out in the hills tending their sheep, darkness suddenly fell upon them. They were terrified and promised God that if they survived they would change and become better men. Soon afterwards, the sun came out again, and legend has it they hugged each other and never fought nor quarreled again for the rest of their lives.

—NATIONAL MUSEUM OF THE FAROE ISLANDS

H aving experienced the power of an eclipse with full warning of what to expect, I can believe its power to change lives when wholly unexpected. The eclipse this time, on March 20, 2015, will be far from a surprise to the 50,000 people who call these islands home. Over the past three days, an additional 11,000 eclipse-chasers have descended upon the islands; each is eager to see what was once a subject of terror, but has now become one of tourism. The lone total solar eclipse of 2015 has not made it easy for the ever-growing throngs of "coronaphiles" to reach their moment in the Moon's shadow. Totality touches land in only two places this time, the Faroe Islands at latitude 62° north, halfway between Scotland and Iceland, and the even more remote islands of Svalbard, 80° north latitude off the coast of northern Norway (total population: 1,200, not including the polar bears).

Based on the latest weather broadcast from Denmark—
the only one I'm getting with English subtitles—the om-
nipresent clouds are forecast to break sometime around
the moment of totality, give or take an hour. I'm told the
chances are better in the northwestern part of the islands,
the direction from which the winds blow, so I have positioned myself on the literal edge of the archipelago as far
north and west as one can go. Thus I find myself sitting
on a wooden bench built to stare out to sea along dramatic
thousand-foot-tall volcanic cliffs teaming with seabirds.
The sound of the distant surf is faint below me and the
sense of being at World's End profound—it's as if here the
Creator ran out of rock and simply quit. Beyond my perch,
nothing more than a few jagged islands dot the northwestern horizon, while beyond them is the open sea.

For most eclipse-chasers, the goal in picking a spot from
which to view an eclipse is to optimize the combination
of clear skies and proximity to the central line of totality.
Being as close to the central line as possible maximizes the
minutes, and even seconds, of darkness. But in the Faroe
Islands, a land of tall mountains and deep valleys, it's crucial to find a spot where the morning Sun is guaranteed
to be above the surrounding hills while eclipsed. Michael
Zeiler, an expert in systems of graphical information, has
created gorgeous maps of the Faroes revealing the sunlight
and shadow across the landscape at the moment of totality. I've seen them everywhere on the islands these past few
days; locals and tourists alike pore over them debating the

perfect spot from which to see the Sun. I've found my sunny spot at the top of a sheep meadow overlooking the tiny town of Gásadalur on the western island of Vágar. Whether it will be free of clouds is beyond the power of Zeiler's maps to promise.

It's an article of faith among eclipse-chasers that the ideal eclipse-viewing spots are those with access to easy mobility so that viewers can quickly drive in one direction or another depending on the clouds. My meadow, unfortunately, sits at the end of a one-kilometer-long, one-lane tunnel through the surrounding mountains. Quick movement is out of the question. Besides, the weather changes so quickly here that expert eclipse-chaser (and psychologist) Dr. Kate Russo claims the best strategy is to find your place and stick with it: "This eclipse will be for those with nerves of steel," she warned after spending a month in these islands, watching the changing weather each day at the time totality is forecast to occur.

As I wait, I look over my shoulder to see yet another squall come ashore; it drenches me, and at two hours until the eclipse begins, I remind myself: nerves of steel. It took me more than twenty-five hours of nonstop air travel on progressively smaller airlines to reach these islands and then drive my rental car (one of the last available) to the house of my host, Lis Mortensen. Mortensen is a curator from Jarðfeingi, the Earth and Energy Directorate, with exhibitions at the local National Museum in Torshavn, the capital of the Faroe Islands. She has created an exhibit on solar eclipses, their causes, and local history (the Faroe Islands are their own country, but officially also part of the Kingdom of Denmark). "I felt it was important and that people would want to come learn about the eclipse," Mortensen told me. "Now that the eclipse has gotten close, it's set attendance records. Everyone wants to know what they'll see and where to go to see it."

It's important to remember how recent a change this is. For all but the past few hundred years of human history, solar eclipses were seen only by those who happened to live within totality's path. The Faroe Islands are no exception. When a total solar eclipse crossed these islands in 1954, the population was still largely isolated from the rest of the world. Radio had only recently come to the islands, and if not for the work of Niels Pauli Holm, a Faroese ophthalmologist, no one would have known what was about to happen and how to observe it. As an expression of the uncertainty and confusion leading up to the 1954 eclipse, Mortensen's exhibit quotes the experience of a young Faroes girl at the time: "I went home and my mother asked me to collect the clothes from the clothes-line. She was afraid that the clothes would burn. People were talking about total destruction, and I remember that people frequently visiting us talked about it. We were asked: 'What are you going to do?' We didn't have any answer to this, looking uneasily at each other. . . . [The old people] were afraid the houses would be destroyed."

The mood is noticeably different now than it would have been then. The streets are filled with tourists, and restaurants and bars all over town feature signs offering limited-edition solar eclipse beer and special fish burgers with homemade tartar sauce. These signs are in English and Dutch as well as Faroese, a language related to Icelandic but spoken only by the people who live here. It is one of many languages in danger of disappearing in our global Internet age, and this eclipse has posed a problem, as there are no Faroese words for a number of the phenomena associated with the eclipse. Wary of letting the language adopt too many English (or worse, Danish) words and losing its heritage altogether, the locals search for analogous Icelandic words to express what we've all come to see: *Sólarmyrking*, the solar eclipse.

Sitting on my bench in this meadow, I am astounded by the number of languages I hear from the eclipse-chasers: Greek, German, French, and more that I can't identify. The Greek eclipse-chasers comment that after a week in the islands I am the first American they have heard. The joy they show seems to indicate that I've given them an advantage in some sort of nationality-bingo game to which I am not privy. My twenty-five hours of traveling also appear to have set a new record for this crowd. But once you've seen totality, traveling halfway around the globe just isn't too far to see it again.

The very first person to travel to see a total solar eclipse—the world's first eclipse-chaser—appears to be Monsieur le Chevalier de Louville, a member of the Royal Academy of Sciences at Paris. He traveled to London to see the total solar eclipse of April 22, 1715, that was predicted by Edmond Halley. Halley was famous for using his friend Isaac Newton's laws of motion to predict the courses of the Moon and comets as well as the dates, times, and places of eclipses both future and past. According to his calculations, this would be the first eclipse to pass over London in 575 years. As one of the first to have its exact time and location predicted (not counting the stories of Miletus of Thales 2,000 years before), odds are this was the first one for which an avid enthusiast could make definite travel plans.

At that time, before the invention of the telescopes and cameras that coronaphiles are setting up around me, the most useful instrument the public had at their disposal for recording an eclipse was the pendulum clock. Halley made use of this fact to refine his orbit for the Moon and perfect future eclipse predictions. In the weeks before the eclipse, Halley sent out broadsides to be posted across the country asking for anyone who could see totality to record its duration using their clock and let him know their results. The map of those who saw totality, as well as the time and duration of its occurrence across the countryside, would determine the exact position and motion of the Moon's shadow, and thus its path around the Earth.

The public was asked to take part in an almost identical act of citizen-science 210 years later in New York City. On the morning of January 24, 1925, a total solar eclipse was predicted to sweep across the state of New York, with the southern edge of its band splitting the island of Manhattan into those who could see totality and the solar corona and those who couldn't. Astronomers from several northeastern universities, including Yale, Princeton, and Cornell, sought to use data from the eclipse to refine their calculations for the size and orbit of the Moon; the data would also enable them to solve some as yet unexplained discrepancies in the time and duration of recent eclipses. Articles filled the local newspapers urging the public to go out and witness the celestial event from their rooftops, from street corners—from wherever they had a view of the Sun—and report back what they saw. The headline and story in the *New York Times* for the morning of January 20 read:

SCIENTISTS ASK AID OF LAITY IN ECLIPSE
Public's Observations Along Edge of Shadow
Counted On for Important Data.
MAY SOLVE MOON'S SHIFTS

If the weather is clear, one of the features of greatest interest for Manhattan will be the determination of the exact line which separates the total eclipse from the partial eclipse. This line is expected to occur somewhere between 110th Street and Seventy-second Street. An observer north of the line will see everything—the complete blackening of the sun's disk, the piercing through of the brighter stars and planets, the thin red rim about the sun, the delicate tree-like scarlet "prominences" outside on the red rim and the pearl-tinted lines of the corona extending in all directions away from the sun.

Amateur photographers and observers in Manhattan will have a chance to help clear up one of the most
difficult scientific questions about the eclipse. By fixing
the exact line which separates the total eclipse from the
partial eclipse, they can establish the exact diameter of
the moon and the exact course of the moon. . . .

While the observer is surer of seeing all the heavenly sights by taking a position north of 110th Street,
he will have the excitement of engaging in cosmic de-
tective work if he stays in the doubtful zone between
Seventy-second Street and 110th Street and studies the
shadow effects from his rooftop.

The article claimed that the edge of totality in a solar
eclipse had not crossed a great population center like New
York City since the invention of the camera for astronomical use. Never before could so many people with so many
cameras photograph an eclipse from along totality's edge,
capturing in the process a phenomenon that, though seen
innumerable times before, had never been named. Whether
as a result of the wealth on display in Manhattan's shops, or
merely a reflection of the roaring excess of the 1920s itself,
the name they gave was perfect and continues to be used to
describe what for me is the highlight of every eclipse: the
diamond ring.

Witnesses that day remarked in great numbers on the
incredible beauty of that final instant before the Sun was
totally extinguished. At that moment, one last ray of the

FIGURE 6.1. Illustration from the January 21, 1925, *New York Times* to prepare readers for the expected view across Manhattan on eclipse day. (Image courtesy USC Libraries / Corbis)

solar disk would have shone down a lunar valley along its limb and produced a brilliant pinprick of light set in the luminous ring of the just-emerging corona. A headline in the *New York Times* three days after the eclipse read: **"THAT 'DIAMOND RING' IN THE SUN'S ECLIPSE: A Remarkable Photograph Taken at Saugerties Seems to Prove it No Illusion."**

Each total eclipse displays two such rings, one at totality's start, the other at its conclusion; to my mind the second always seems most beautiful as a final punctuation on the spectacle just finished. Of the roughly 6 million people living in New York City in 1925, those in Manhattan right along the southern edge of the shadow would have seen a totality of no more than an instant—consisting of a single diamond ring—the jewel-like gleam of the Sun never fully hidden.[*] Because winters are cold in New York and the forecast was for snow the day of totality, an unusual plan was put in motion for a fleet of twenty-five airplanes (including a dirigible) to take to the sky with cameras and other instruments. They were to record the moments of totality and broadcast their results through a constant stream of radio commentary to the public below. A *Times* reporter captured the drama of the largest formation of airplanes to fly across New York since the end of the Great War only six years before:

> As the machines winged toward their destinations, the shadow [of the Moon] grew larger, while Major Hensley, his lips only an inch or two from the microphone, told the millions within radio distance what was going on. . . . Looking from the cockpit of [his] airplane, the observers saw the shadows deepening over Long Island Sound and the Connecticut shore. Far off on the horizon, at the extreme northerly edge of the shadow, a play of soft green, purple and deep blue light could be seen. . . . Then came totality and with it the darkness of night. . . . Higher and a little to the right the planets Mercury, Venus and Jupiter glowed with a soft blue light as they burst into prominence with the dimming

---

[*] In 2017, the residents of St. Louis and Kansas City, Missouri, will get the opportunity to see this same phenomenon in exactly the same way, as each city is split in half by the eclipse's path.

of the greater luminary. . . . The pilots of the machines, though their minds were intent on their jobs, found times to gaze on the spectacle. The scientists in the cockpits were enraptured. They saw the eclipse under conditions that no others had ever experienced.

The novelty of the eclipse was also communicated to the public through moving pictures taken from Yale Observatory in New Haven, Connecticut, and then sent by airplane to New York City so that they could be playing in theaters on Broadway by that afternoon. Airplanes play an even larger role in eclipse-chasing today. As I sit at the edge of the Faroe Islands I can see blue sky coming over the horizon, and can only pray that it gets here in time. Above me, however, are thirty specially chartered airplanes full of passengers with no need to worry about what the weather will bring. Three of them are Boeing 737 jet airliners that took off from Iceland to be here at totality, while three private, nearly supersonic jets have flown from Paris and Geneva, Switzerland. At the moment of totality, they will literally turn to chase the Moon at nine-tenths the speed of sound: fast enough to prolong totality from the two-minute, twenty-second event I hope to see from land to almost four minutes.

They are not the first to prolong totality in this way. The record for the longest duration of darkness is held by the supersonic Concorde, which as a mere prototype in 1973 was chartered by an international group of scientists to streak across Africa during an eclipse. From an altitude of 55,000 feet, the sky was black and the curvature of the Earth clearly visible, as was the shadow of the Moon beneath them. The Concorde traveled at a speed of almost 1,300 miles per hour, about twice the speed of sound and the same as the speed of the Moon's shadow moving across the Earth, stretching the view from the ground—an exceptionally long seven minutes—to an astounding seventy-four minutes of totality in the sky.

Older coronaphiles tell me that the 1970s was the decade when commercial eclipse-chasing really began. The first public cruise ship chartered to see a total eclipse of the Sun occurred in 1972 (nine hundred miles off the coast of New York in the western Atlantic), the same year as the release of the Carly Simon song "You're So Vain," in which the unnamed subject of the song flew his "Learjet up to Nova Scotia to see the total eclipse of the sun." The first commercial flight for amateur solar eclipse-chasers was an Ansett Airlines flight chartered out of Perth, Australia, in 1974: all the seats were removed from the left side of the plane so photographers could set up cameras and telescopes to look out the small windows.

Public interest in eclipse-chasing has grown exponentially in the decades since. In 2001, Doug Duncan, an astronomer, educator, and longtime eclipse-chaser, planned to charter the Concorde to reproduce the extreme-duration 1973 eclipse, but this time for members of the general public who could afford the cost. At $10,000 a seat, it was only marginally more expensive than a typical Concorde flight, but given the position of the Sun low in the sky, passengers would see totality perfectly framed in the Concorde's tiny windows for the entire hour-and-a-half flight across the Atlantic. Unfortunately, the Concorde that he was in the process of chartering was the one that crashed upon take-off in 2000, after which the fleet was grounded, never to fly again.

Chartering cruise ships for eclipse-chasing was no less difficult. "They wouldn't give me the time of day for two years," Duncan told me when he tried to find a Mediterranean cruise line that was willing to alter course by a hundred miles to intersect totality off the Greek Islands in 2006. "Then finally, a year before the eclipse, they came back and said to me, 'We don't understand why, but a lot of people want to be on a ship going to where you want to go. So we'll let you have a third of our ship. Send us a non-refundable deposit of $100,000 by the end of the month and it is yours.'" Duncan had to take all of his savings and another mortgage on his house, but he paid the deposit and nearly filled the ship: "I ended up taking 402 people," he said. "I hired 10 astronomers to speak, and I ran a kids program for 50 people, and all the kids were flying kites off the back of the cruise ship. It was glorious." Today, every eclipse that crosses any sizable body of water is almost guaranteed to pass over a cruise ship carrying an array of expert speakers for the crowds, including astronauts, astronomers, and scientific authors. In fact, there are nine ships in and around the Faroe Islands today for this eclipse.

The most dramatic solar eclipse I've ever seen was on just such an eclipse cruise across the Atlantic in the fall of 2013. It was from the deck of a four-masted luxury sailing ship, the *Star Flyer*, sailing from Spain to Barbados, and I was a speaker. Nearly the entire ship had been booked, and so the cruise line agreed to alter course and intercept the Moon's shadow for the forty-two seconds that totality would be visible from off the coast of Africa 20 degrees north of the equator. It took a full week of sailing out of the Canary Islands under nothing more than billowing white clouds and baby-blue skies. In an almost eerie counterexample to what I am experiencing today, the day of totality was the only day that dawned cloudy. But with masterful sailing by our captain, at the last moment before totality we managed to reach the lone break in the clouds. The Sun disappeared behind the Moon at the exact instant we crested a wave and broke free of the gloom. Clutching the rigging with one hand and my hat with the other, I was enthralled, the moment made all the more special by its brevity and a horizon circled with storm clouds and rain in every direction but the one that mattered.

Back in the Faroes, one of the people in the air overhead is Bárður Eklund from the Visit Faroe Islands tourism board. He hangs from the open door of a helicopter photographing the Moon's shadow racing across the cloud-tops. Dr. Kate Russo, the eclipse-chasing psychologist, has been working with the tourism board and others on the islands for over a year, using her experience to help prepare the community for what is about to happen here.

Eclipse-chasing scientists haven't always had a very good record of sharing the beauty of what they have come to see with the people who actually live there. The Faroes eclipse of 1954 took place during the start of the Cold War, when the islands were an early warning station for the North Atlantic Treaty Organization (NATO). A small team of American scientists traveled here to study the eclipse, and the only record they left is a tiny plate on a stone in the village of Lopra with this inscription: "Solar Eclipse Expedition 30 June 1954 US Air Force." That is par for the course for many of the solar eclipse expeditions stretching back through the 1800s. Local populations were viewed as a source of free labor (at best), and a potential source of danger and theft (at worse).

Alex Soojung-Kim Pang, a former deputy editor of the *Encyclopaedia Britannica*, has researched the intersection of Victorian-era eclipse expeditions, tourism, and the people whom eclipse-chasers encountered. He describes the racist attitude held by many Western scientists when on expedition, writing that, "before an eclipse, [local] crowds were merely troublesome, but during an eclipse they were far more dangerous: stirred up by jealous priests, shackled by ancient superstitions, constitutionally incapable of the same kinds of self-control on which Europeans prided themselves, they threatened to revert to savagery under the enormous emotional pressures of totality." The solar

astronomer Norman Lockyer, referring to his 1871 eclipse
expedition, reported that his observations could have been
ruined by "the smoke of [Hindu] sacrificial fires, . . . if there
had not been a strong force of military and police present to
extinguish them." Moreover, he wrote, "in Egypt, in 1882,
without the protection of soldiers, a crowd of Egyptians
would have invaded the camp."

This was the era that saw the beginning of a professional traveling class. This was the new class of tourists who
could afford months, if not years, on the road and who were
well moneyed, well educated, and well read. They were en-
couraged to read up on the geology, flora, and fauna (and
occasionally the people and customs) of their exotic destinations. The same railroads and steamships that made
worldwide travel possible—as well as the rise of institutions
like tourist hotels, travel agencies, and travel literature to
cater to Victorian tourists—were indispensable for the new
eclipse expeditions, and the members of those expeditions
were often drawn from the same well-connected leisure
class. The attitudes of the scientists toward the local inhabitants they encountered therefore mirrored the attitudes
as a whole of the Victorian Age: "It is not at all probable
that one of the dusky lookers-on at our preparations had a
remote idea of the approaching phenomenon, and certainly
not of the objects of our arrangements. . . . No effort could
have given them the slightest comprehension of the causes
of the unusual darkness, nor why the white man should
come so far to look at it."

Having seen multiple eclipses myself, it is my fervent belief, as it is others', that eclipses should be enjoyed by everyone fortunate enough to be in the path of totality, not just the scientists or the dedicated tourists who have come to see them. Today, Dr. Russo is leading this charge to share totality's beauty beyond the crowds of already excited eclipse-chasers. Back in 2013, two years before the eclipse that has brought us both here, she arrived in the islands to work with local officials and spread the word about what would occur: "I really thought it was important to be a part of the community," she told me, "being here, building it up, sharing it with the community, not just being an eclipse-chaser, coming in, seeing it, and going."

She had experienced what it was like to be a part of the local community during an eclipse in 2012, when totality touched her native Queensland on the northeastern coast of Australia. Living abroad, she had returned there to promote her book *Total Addiction*, about the psychology of seeing an eclipse. "I had gotten there a month before and I was doing a survey of people before and after the eclipse," she said. "There was nobody on the ground in North Queensland who had actually seen an eclipse before." People who lived there were talking about leaving, not even staying to see the eclipse because they were so worried about the influx of tourists. "Oh no," Russo thought. "This is coming to your community; you need to be here." She started going to markets, community groups, and schools to give talks, and she made herself available for radio and TV interviews as someone who had seen what an eclipse was like. "And the more I did," she said, "the more interest there was, because once you've seen an eclipse and start talking about it you can't help but become excited."

The great fear for Russo, as it is for me, is that for those who talk about eclipses without ever seeing one—the local officials, the radio hosts, and the TV reporters that flood the airwaves in every metropolitan community before totality—the passion with which the experience grabs you as a physical thing is difficult to understand before the fact. They can leave the public with a distorted view of what an eclipse will be. At best, they may suggest it is some sort of scientific novelty, an educational event that is worth seeing if you are free at the time. At worst, they may leave their viewers with the impression that eclipses are for new-age oddballs, and that making the effort to see an eclipse is something only slightly demented people do. This is understandable; if you haven't seen one yourself, you can't help but not understand the experience.

It's for that reason that Russo now travels to eclipse communities to share her knowledge of what the experience is like. A year before the eclipse in the Faroes, she and Geoff Sims, an eclipse-chaser with a PhD in meteorology from Australia, set up a citizen-science project asking local Faroese to photograph the sky where they live at exactly 9:40 a.m. (the time of the eclipse) every day for a month to compare with local weather statistics, so that they would be able to gauge the chances of clear skies at different points across the islands on eclipse day. In February 2015, a month before totality, Russo returned to the islands to work with the tourism office as well as local media, schools, museums, businesses, and artists. Their goal was to develop and disseminate information for tourists and locals alike on where to go, what to look for, how to see it safely, and how to ensure that as many people as possible could share the moment together. "A year ago," she said, "even when they were thinking that maybe no more than 5,000 people would be traveling to the islands to see the eclipse, that still left 50,000 people here who needed to know what was happening."

In my week before the eclipse, I heard about her efforts firsthand on the local radio. I learned about one community where a hospital scheduled no surgeries so staff could pop outside for twenty minutes to see the lead-up to totality. Elsewhere, schools made plans to let students stay home with their families to enjoy the experience together, while in other communities, the local schools became the center of the community event.

Russo became interested in eclipse-chasing in 1999 when she witnessed her first solar eclipse on the coast of France, having traveled there by bus from Belfast to see it. At that same moment, I, too, had been experiencing my first total eclipse, but a little farther along totality's path, in central Hungary. For both of us, it was a profoundly moving experience. She calls us "Saros Siblings," a term she's coined for those of us who have shared in this new experience during the same eclipse. It's no accident that the first hotel room in the Faroes was booked for this eclipse back in 1999 almost immediately after that event. Evidently, another Saros Sibling of ours felt the need to see the corona again and didn't want to miss out on a chance to get a room. In 2017, there will be at least 9 million new Saros Siblings as people flock to totality's path in the United States.

B ack on my clifftop, I'm still sitting on my chosen wooden bench. There are now more than a dozen of us. We've set up our cameras and are swapping stories of eclipses seen and missed because of weather. It's no one's first eclipse: some have seen five or eight; one gentleman has even seen seventeen total eclipses in his travels around the world. Far from satiating their desire, like checking an item off a bucket list, it has left each person wanting to see more. What will it be like in the United States when instead of a dozen we have ten thousand viewers all gathered in one location?

The blue sky is suddenly upon us, and as the clouds part overhead, we can see the eclipse is already underway. The excitement is palpable; it's what makes the experience so memorable for each person who sees one.

I've spoken to many coronaphiles over the years, and the stories they tell of their first time seeing totality reveal the sense of awe that is found in the shadow of the Moon. Eclipse-chaser and expert photographer Geoff Sims saw his first total solar eclipse in 2002 in southern Australia. He had read all about eclipses: including what to expect and how to photograph them. He piled all of his equipment in a car that he drove for three days across the continent to reach a point near the center of totality. What he experienced there was more than he'd expected: "On face value," he once told me, "it is everything that you read about, but the excitement during the lead up, the chills that you get when the Sun gets covered and you realize totality's eminent, that kind of excitement I could never have imagined. Then when you see the corona, it just blew me away. Because you can't describe exactly how that appears in the sky, and photos just don't do it justice. You can never anticipate what it will really be like for the light to just disappear so quickly." His photographic work on eclipses has now taken him all over the world, and it was thanks to his efforts in scouting locations on the Faroe Islands a year ago that I've been able to meet so many people here.

David Makepeace is a filmmaker from Toronto, Canada, who, like many of the growing number of eclipse-chasers worldwide, is in no way a scientist. His first total solar eclipse was in 1991 in Baja California, Mexico. A girlfriend invited him down to see the eclipse. "I had taken an astronomy course at the University of Toronto so I had some kind of basic understanding of what would happen," he shared. "But I went there primarily to see her; the eclipse on the beach would be secondary. Then we saw it and it totally blew my socks off. I was silent for two days afterward, sitting on the beach staring off at the Sea of Cortez wondering about my existence, wondering what I was doing here on this big rock flying through space." Today he works on films and planetarium programs to share this experience with the public.

I look up and see that my sky overhead is now a race between scattered clouds and the Moon. One minute it's clear, the next it's cloudy. Each time the Sun reappears, it is a little farther gone and the colors even stranger. The Sun is now so small that the shadows are sharper, cast by a single white spotlight.

Everything is happening so fast. The light fails, the temperature drops, and suddenly new clouds form over the mountains around us. Just thirty seconds before totality begins, clouds seal the sky shut, and then it goes black. Each one of us is no more than a silhouette under a sky now darker than any day I have ever experienced. The clock begins: two minutes is all we have.

All we need is a momentary break in the clouds anytime in the next two minutes, and we will easily see the corona with our now dark-adapted eyes. Even for a cloudy day, this darkness is unnatural. I can understand the fear of those early quarrelsome brothers in the ancient Norse story. It's absolutely obvious that above these clouds something strange is happening.

FIGURE 6.2. Clouds sweep in during the final moments before totality as the crescent Sun shines down on me from above the volcanic cliffs of the Faroe Islands in 2015. (Photo by the author)

We wait. We look all around at the unbroken blanket overhead. Just a single break is all we need. One minute gone. Sixty seconds left. Is it slightly clearer over there? No?

The horizon grows light, the Moon's shadow is leaving . . . and then a second dawn breaks as the clouds everywhere grow bright and drift apart. Once more we are in Sun.

I missed it.

I fold up my tripod and put away my camera to occupy my thoughts for just a little while. We all joke that we'll see each other next time in Indonesia, or the United States. It helps dampen the disappointment.

Later, at the dinner table of Lis and her husband, Andras, with their family gathered from across the islands, we tell stories of what we'd seen, and in time, we laugh. We talk about who was clouded out and who wasn't. On the roof of the hotel the staff saw totality, while the tour group staying there, who had traveled to a special location for the event, did not. As we talk, we all agree that the sense of having shared in something awe-inspiring in the darkness truly made this day unique for everyone (even under the clouds).

Súsanna Sørensen of the Visit Faroe Islands tourist board told me later about her own experiences, after all the work of helping others to see the eclipse: "We had invited my family and my husband's family to early breakfast at our house. It was a very special morning, with an exciting atmosphere prior to the eclipse. The weather was not too good and it was fantastic when we saw first contact; it was finally here and we could see at least parts of it. Breakfast had to wait and we all went out on our balcony. It was a very strange feeling when totality began, how the light disappeared, like turning off a switch. It was a really beautiful light with a yellow rim at the horizon." Though, like my group, she was clouded out for totality itself, she was philosophical about it: "There was a small hole in the clouds where you could see the blue sky, and we knew that somebody else probably saw totality through that hole."

For those who did, and even those who didn't, we all saw and shared something that day, and we felt lucky to have experienced what we did. A month later I received an email from Súsanna, who told me that, even now, "it has been the talk of every social gathering I have been to since, even this weekend, when we were out with friends. The dramatic light is something that everybody talks about and how it made us all feel small and at the same time part of something bigger."

Like so many others who witnessed that eclipse, she is now thinking about where she can go next to see another. In some sense, the answer to that question is easy. Unlike many other natural spectacles, there is no question about exactly what date the next total solar eclipse will begin, or even the exact moment. All that is in doubt is if we will be there to see it, and who will be fortunate enough to be there with us to share the experience.

When I had once turned my eyes on the moon encircled by the glorious corona, then on the novel and grand spectacle presented by the surrounding landscape . . . I mentally registered a vow, that, if a future opportunity ever presented itself for my observing a total eclipse, I would give up all idea of making astronomical observations, and devote myself to that full enjoyment of the spectacle which can only be obtained by the mere gazer.

—WARREN DE LA RUE, 1862, FIRST PERSON
TO PHOTOGRAPH A TOTAL SOLAR ECLIPSE

CHAPTER 7

The Great American
Eclipse and Beyond

T otality's eerie light bathes the ring of massive mono-
liths as robed figures chant and burn their sacrificial
offerings. Around them, the gathered throngs raise their
voices in joy and smile for the cameras of the CBS Evening News. It is 1979, and the scene is a bizarre roadside
re-creation of Stonehenge overlooking the Columbia River
in western Washington State. Neo-pagans and curious onlookers have amassed to witness the rare event unfold overhead, and this being Washington (and the 1970s), it's clear
that not all the smiles and good vibrations are due solely to
the solar eclipse.

Back in the studio, Walter Cronkite tells us there will not
be another total solar eclipse to touch the continental United
States this century. Not until the far-off date of 2017 will totality once more be so visible to so many on this continent.

I've been waiting to see this eclipse ever since.

The February 26, 1979, eclipse only touched a corner
of the United States before swinging up into western Canada, and not even half the population of the United States
today was alive for it. A continent-spanning eclipse hasn't
occurred in the United States since 1918, almost 100 years
before what is being called the Great American Eclipse of
2017. While this may seem like a long time, consider that
when Francis Baily saw his first totality in 1842, Western
Europe had gone 109 years without seeing the solar corona
at all. Halley's eclipse in 1715 marked the end of a 500-year
drought for England. Australia is currently in the midst of
seeing eight total solar eclipses in a period of 64 years (1974
to 2038), but only after having seen none for the previous
52 years. Similarly, when the current American drought of
thirty-eight years without a total solar eclipse ends in 2017,
it will mark the beginning of a new thirty-eight-year period in which Americans will get to see five (in 2017, 2024,
2044, 2045, and 2052).

While, on average, any one spot on Earth will experience solar totality every 370 years, remote Baker Island
in the Pacific Ocean is in the midst of a 3,000-year gap
between totalities. Over time, these cycles of bounty and
absence come and go, and every place on Earth is crossed
eventually. For human beings, with our limited lives and
limited means of travel, these vagaries of celestial alignment mean the majority of people on Earth have never
seen a total solar eclipse.

THE 2017 SOLAR ECLIPSE

The first total solar eclipse most Americans will have ever seen begins the morning of Monday, August 21, 2017. It will begin two seconds before 10:16 a.m. Pacific Daylight Time (PDT). At that moment, the dark shadow of the Moon touches the Pacific Coast at Yaquina Head lighthouse outside the coastal town of Newport, Oregon. There is no doubt about this. Astronomers have a bad reputation when it comes to predicting amazing sights for the public. Too many "Comets-of-the-Century" turn into faint fuzzy duds that disappoint in the darkness. Too many meteor "storms" wind up being no more than a drizzle once you've woken the family at 2:00 a.m. But this eclipse is happening, in the middle of the day, exactly on time, and in exactly the places that are predicted. It is as certain as the sunrise.

The only question is a matter of clouds, and even those can be forecast with some certainty. The region with the best chance of clear skies all along the path of totality on that date is eastern Oregon, which the Moon's shadow reaches at 10:19 and 36 seconds PDT as it crosses the Cascade Range and flies eastward at a speed of 2,265 miles per hour. The shadow then crosses into Idaho and the southern tip of Montana, where no roads lead into the Beartooth Mountains.

At 11:34:56 a.m. Mountain Daylight Time (MDT), the shadow reaches Jackson Hole Airport in the middle of Wyoming's Grand Teton National Park. This is one of two national parks totality touches, and the only one in the high mountain air of the western United States. Summer storms can be an issue, though. The weather for the week of August 21, 2014 (exactly three years before the eclipse), included rain and sleet; a year later, on August 21, 2015, it was mostly clear and sunny. The eclipse crosses Wyoming diagonally almost entirely along the line of Highway 26. For those looking to avoid any clouds and reach potentially clear skies in their cars, this could be the ideal racetrack for whichever direction it's necessary to drive.

The lunar shadow then sweeps across the plains, reaching the border of Nebraska at 11:46 a.m., and the northeast corner of Kansas at 1:02 p.m. Central Daylight Time (CDT). Twenty-two minutes later, totality engulfs Kansas City, Missouri—the largest city yet along the path—right after lunchtime. A little farther east, at the University of Missouri in the town of Columbia, residents are expected to gather in the 71,000-seat Faurot Field football stadium. It will likely be the largest single gathering of spectators to see totality on the continent this day.

Totality reaches the Mississippi River south of St. Louis at 1:17:08 CDT. People living in Missouri's largest city will see the Sun disappear above their homes, but only if they live south and west of the line between Creve Coeur Park and the Missouri Botanical Gardens. For those wishing to see totality from Forest Park, home of the famous 1904 St. Louis World's Fair, only the extreme southwestern corner of the park is within the path, and totality there lasts no more than 20 seconds.

At this point, onlookers all along the midline of the path have had at least two minutes of totality, but the residents of Carbondale, Illinois, will get to experience the eclipse's greatest duration, two minutes and forty seconds. These folks are luckier still, as the very next total eclipse to touch the continent will sweep over their houses in 2024, allowing them to see two eclipses in seven years from their very own doorsteps.

From Illinois, the path leads over Kentucky and then crosses into Nashville, Tennessee—the largest city completely within the shadow of the Moon at 1:27:28 CDT. As totality leaves Tennessee, it crosses the southern half of Great Smoky Mountains National Park, the most visited national park in the United States. In 2010, 9.4 million visitors came to this park, more than twice the number that visited the Grand Canyon that year. The highlight of the park is Cades Cove, a spectacularly beautiful valley set amid hardwood forests and rushing streams. Totality reaches here at 2:34:20 Eastern Daylight Time (EDT) and is visible in the park from much, but not all, of Newfound Gap Road as it crosses from Tennessee into North Carolina. The weather here can be hazy, though, especially on a summer's afternoon. There's a reason they call them the *smoky* mountains.

At 2:47 EDT, one hour and thirty-three minutes after coming ashore on the Pacific Coast, the Moon's shadow crosses out into the Atlantic just north of Charleston, South Carolina. In that time it will have touched thirteen states, five state capitals, and 9.7 million residents, not counting the millions that will travel into the path to see it. For that one day, every man, woman, and child in North America will share in a phenomenon of wonder and joy. Though totality will take only ninety-three minutes to cross from sea to shining sea, during that time everyone on the continent will be within at least the partial shadow of the Moon and experience the eclipse together.

But not all views are equal.

Although the entire continent will witness at least a partial eclipse, the real show is within the zone of total darkness. You must get into the *umbra*, the darkest part of the Moon's shadow where the disk of the Sun is totally covered. That is where the eclipse is total. Those in totality's path see the diamond ring, Baily's beads, the corona, and any prominences visible on the solar limb. Day becomes night and the brighter stars and planets, such as Venus and Mars, become visible near noon. You will understand the true meaning of awe.

While those near the midline of totality will see the longest darkness, those near the edge—but still just within the umbra—have the chance to see more edge effects, such as Baily's beads and the deep red chromosphere (including prominences) that appear as the Sun just skirts behind the lunar edge. Those barely outside totality's path may still see some Baily's beads (if they are within just a few miles of the band); but, although the day will grow dark and colors will change, day will not become night, and no stars or planets will become visible. Here, outside the path of totality, the Great American Eclipse will be only partial, and special eclipse glasses will be needed for the entire event.

Make no mistake: The difference between being inside and outside the path of totality *is* the literal difference between day and night.

A GUIDE TO SAFELY VIEWING A SOLAR ECLIPSE

I saw the 1979 eclipse, my first, in complete safety at home: in a darkened room, with all the windows covered, watching it on TV. I missed experiencing the greatest awe-inspiring wonder that Nature has to offer because our local school officials—who felt that eclipse-watching was too dangerous for children and did not want to be held liable in the event of an injury—had issued overly cautious warnings. Regardless of whether those school administrators were right to be wary, not a year has gone by since that I haven't felt cheated out of a life-changing experience.

Observing a solar eclipse *can* be dangerous if proper precautions are not taken. In that regard, viewing a solar eclipse is no different from many other sporting activities in which families and schools regularly take part. But unlike bicycle helmets, football pads, or lifejackets, the equipment needed to safely view a solar eclipse costs no more than a dollar or two, and you may even have it sitting around your house already.

First, you may be wondering what the potential harm is. The un-eclipsed Sun is visible every day, yet we don't warn children to stay indoors on sunny days. While it is true that the Sun doesn't emit any special rays during an eclipse, how we behave toward the eclipsed Sun does change. On a typical day, very few people willingly stare at the Sun for a prolonged period of time. Our eyes and brains know it isn't good for us to shine that much light through the lenses of our eyes and focus it on our retinas, so we instinctively look away.

The concern for health officials during an eclipse is that people will stare at the Sun, particularly close to totality when the Sun may look like a crescent. People want to see this phenomenon, and so they will look at a particular sharp feature longer than they should, placing its image on a single part of the retina. Since there are no pain receptors in the retina, the damage that takes place as it burns produces no discomfort. For all its painlessness, the damage can be permanent. And particularly for young children with clear pupils that have not become dulled with age, the increased amount of light on the retina does make for more danger compared with adults.

The solution to this problem is simple: never look directly at the partially eclipsed Sun without proper eye protection. Either look through specially designed, filtered glasses, or project an image of the Sun on a screen, using a pinhole projector made from items found around the house or naturally outdoors. These two methods of seeing the partial phase of a solar eclipse are cheap, safe, and in the case of finding naturally occurring pinhole-projections, an enormous amount of fun.

Filters: The most widely available device for viewing the partial phase of a solar eclipse is simple, commercially available plastic or cardboard safety glasses. These glasses are inexpensive, typically $1 or $2, and are available from a number of websites. Local stores usually sell them in places where total eclipses occur. (See the end of this book for licensed suppliers.)

You should be unable to see anything through the lenses of these glasses other than the Sun. The special solar filter in these glasses can be delicate. Any holes or scratches they develop make them instantly worthless and potentially dangerous. Keep them protected. If you have any doubts, hold them up to a lightbulb: if you see any light at all coming through the lenses, throw them away and use another pair.

What not to use: Stay away from people offering to let you look through makeshift items like dark beer bottles, silver candy wrappers, CDs or DVDs, smoked glass, or dark sunglasses. These may make the Sun look dim, possibly even dim enough to look at without discomfort, but they may do nothing to block infrared or ultraviolet radiation that will cause permanent damage to the retina and possibly lead to blindness.

Projection: The cheapest, simplest, and absolutely safest method of watching a partial solar eclipse is to find a

FIGURE 7.1. A small hole in a card, hat, colander (or any other item) will project an image of the partially eclipsed Sun. (Image by the author)

piece of cardboard and poke a small hole in it. Hold this outside and the sunlight passing through the hole produces an image of the Sun wherever the shadow of the cardboard falls. Place a sheet of paper on the ground or on a wall and everyone can see the eclipse progress together. Do not look through the pinhole at the Sun.

One of the most enjoyable parts of watching a partial solar eclipse (or spending time outside in the Sun waiting for totality to begin) is looking around for natural pinhole projectors. You can use leafy trees, woven hats, interlaced fingers, or any other place that tiny holes occur through which the eclipsed Sun shines, casting myriad crescents into their shadows.

Whether using filters or projection, once totality begins with the first diamond ring, you may put the filters and glasses aside and feel free to look at the Sun with the naked eye in complete safety, until the second diamond ring marks totality's end.

PHOTOGRAPHING SOLAR ECLIPSES

My simplest recommendation for photographing a total solar eclipse is: Don't do it! Typically, you will have no more than two minutes of totality. That's 120 seconds. Why spend those precious seconds looking down at your camera's instruction menu, trying to get your camera to focus, or working to get the exposure right? Reread the quote at the opening of this chapter. Even the very first person to ever photograph a total solar eclipse wished he had the chance to see another without the bother of equipment. If this is your first eclipse and you feel you must have a photo, someone else will take one and it will probably be better than yours.

If none of that dissuades you, here are a few tips to keep in mind for taking solar eclipse photos. There are links to specific websites with more detailed information at the end of this book.

1. During the partial phase of the eclipse, do not point your camera at anything you wouldn't look at with the naked eye. You may be able to safely look at the monitor on the back of your digital camera, but your camera lens is focusing the Sun's light on your sensitive camera optics. If you need a filter for your eyes, so does your camera.

2. If you are using a filter for your camera during the partial phase, remember to take it off during totality or you won't capture anything.

3. During totality, the sky darkens enough for the brighter stars to appear. In the days before totality, go outside immediately after sunset and wait until the first, brightest stars appear. This is a fair approximation of how dark it will be. Experiment with taking photos at this time. How do they come out? Do you need a tripod for your camera to successfully take sharp, non-blurry, non-grainy photos under these conditions? If so, this will be another piece of equipment you will need to handle during those precious seconds of totality.

4. It will be dark during totality, but do not use a flash. It will blind everyone around you at exactly the moment they want to see the sky the most. Turn all flashes off.

5. The Full Moon is about as bright as the corona. Try taking photos of the Full Moon to see how long you need for the Full Moon to be properly exposed. Again, does this need a tripod?

6. The Full Moon is also the same size as the Sun and will be the size of the "hole in the sky" during the eclipse.

Practice using your camera to photograph the Full Moon. How big does it appear in your image? Is this worth taking a picture of? Many of the best eclipse photos that show details of the corona and prominences use telephoto lenses with focal lengths of at least 500 millimeters (mm), but they cost thousands of dollars.

FUTURE ECLIPSES

If you miss the total solar eclipse of 2017—or, having seen it, you catch the bug and absolutely must see another—don't worry, there are many more coming. The next total solar eclipse to touch the continental United States is April 8, 2024. On that day in midafternoon, the path of totality starts in Mexico, travels northward into Texas, and then crosses the central United States before passing through eastern Canada. The citizens of southern Illinois in the United States are the lucky ones: they will get to see two total solar eclipses in seven years without having to travel anywhere.

Perhaps the most spectacular views in 2024 will come for those on the US side of Niagara Falls. From the railing beside the water, the totally eclipsed Sun will hang directly above the crashing falls for three minutes and thirty seconds in what is sure to be a wild sensory-overload spectacle of sight and sound—assuming it is clear at that time of year, of course.

For those who like to travel, there are some exotic destinations for future solar eclipses that are tantalizing for the
opportunities they afford. In July 2019, totality will sweep
off the Pacific Ocean and across the Chilean Andes. The
path will cross one of the great astronomical observatories
of the Southern Hemisphere, the La Silla Observatory at
an elevation of 7,900 feet (2,400 meters), operated by the
European Southern Observatory. This is one of the darkest
locations on Earth, and since solar eclipses happen at New
Moon, those who travel here for the eclipse should stay to
see the glories of the southern Milky Way high overhead
throughout the night. It is a view unlike any available to us
in Europe or North America, particularly with regard to the
darkness of the night and the grandeur of our galaxy.

On December 4, 2021, at the height of the Southern
Hemisphere's summer, the totally eclipsed Sun will be visible from Antarctica and no other continent. The Sun and
Moon will glide horizontally no more than about 20 degrees
off the gleaming ice—about the distance spanned by your
thumb and little figure extended at the end of your arm.

Surely one of the most awe-inspiring sights will be the
total eclipse of August 2, 2027, from within the Temple of
Luxor in Egypt. Totality will last over six minutes, the longest for the next thirty years, and hang in the darkened sky
almost at the zenith, directly above the stone pillars and
statues below. For those who can't get to Egypt on that day,
this particular path first touches ground passing precisely
over the Strait of Gibraltar and the Pillars of Hercules (the

TABLE 7.1. Total and Annular Solar Eclipses Worldwide, 2017–2030

Date	UT Time Greatest Eclipse	Type	Saros	Duration Tot./ Ann.	Path of Totality/ Annularity
2017 Feb. 26	14:54:32	Annular	140	00m44s	Pacific, Chile, Argentina, Atlantic, Africa
2017 Aug. 21	18:26:40	Total	145	02m40s	Northern Pacific, US, Southern Atlantic
2019 July 02	19:24:07	Total	127	04m33s	Southern Pacific, Chile, Argentina
2019 Dec. 26	05:18:53	Annular	132	03m39s	Saudi Arabia, India, Sumatra, Borneo
2020 June 21	06:41:15	Annular	137	00m38s	Central Africa, Southern Asia, China, Pacific
2020 Dec. 14	16:14:39	Total	142	02m10s	Southern Pacific, Chile, Argentina, South Atlantic
2021 June 10	10:43:06	Annular	147	03m51s	Northern Canada, Greenland, Russia
2021 Dec. 04	07:34:38	Total	152	01m54s	Antarctica
2023 April 20	04:17:55	Hybrid	129	01m16s	Indonesia, Australia, Papua New Guinea
2023 Oct. 14	18:00:40	Annular	134	05m17s	Western US, Central America, Colombia, Brazil
2024 April 08	18:18:29	Total	139	04m28s	Mexico, Central US, Eastern Canada
2024 Oct. 02	18:46:13	Annular	144	07m25s	Southern Chile, Southern Argentina
2026 Feb. 17	12:13:05	Annular	121	02m20s	Antarctica
2026 Aug. 12	17:47:05	Total	126	02m18s	Arctic, Greenland, Iceland, Spain
2027 Feb. 06	16:00:47	Annular	131	07m51s	Chile, Argentina, Atlantic
2027 Aug. 02	10:07:49	Total	136	06m23s	Morocco, Spain, Algeria, Libya, Egypt, Saudi Arabia, Yemen, Somalia
2028 Jan. 26	15:08:58	Annular	141	10m27s	Ecuador, Peru, Brazil, Suriname, Spain, Portugal
2028 July 22	02:56:39	Total	146	05m10s	Australia, New Zealand
2030 June 01	06:29:13	Annular	128	05m21s	Algeria, Tunisia, Greece, Turkey, Russia, Northern China, Japan
2030 Nov. 25	06:51:37	Total	133	03m44s	Botswana, South Africa, Australia

A hybrid eclipse is one that begins as annular but becomes total (eclipse predictions by Fred Espenak, NASA's GSFC).

Rock of Gibraltar in Spain and Jebel Musa in Morocco). Totality lasts for about four and a half minutes there before continuing over northern Africa to Egypt and across the Red Sea.

The solar eclipse of July 22, 2028, passes directly over Sydney, Australia, affording three and three-quarter minutes of totality, while South Australia is visited by totality just two years later on November 25, 2030.

Lest we forget total lunar eclipses, there is a special beauty to going out under a night sky brightly lit by a Full Moon. What begins with only a few stars visible against the glare of the Moon slowly turns into a thousand points of light as the eclipsed Moon darkens. By the time totality occurs, the sky is ablaze with stars; if totality occurs near the peak of a meteor shower, the dark-red Moon can be surrounded by shooting stars for the duration of totality. I saw this in August 2007 from Grand Teton National Park immediately after the Perseid meteor shower with the Milky Way arching overhead; it was one of the most magical moments I've ever experienced.

For such a night, consider the total lunar eclipse of July 27, 2018. It will occur two weeks before the peak of the Perseid shower, but a few meteors may be visible as the blood-red Moon hangs beside the brightest portion of the Milky Way for witnesses in Europe and Africa through Asia. To find more possibilities, and see the places you'll need to go to follow the path of totality, visit any of the websites listed at the end of this book.

TABLE 7.2. Total Lunar Eclipses Worldwide, 2017–2030

Date	UT Time Greatest Eclipse	Saros	Umbral Mag.	Duration of: Eclipse (Totality)	Visible
2018 Jan. 31	13:31:00	124	1.315	03h23m (01h16m)	Asia, Australia, Pacific, Western N. America
2018 July 27	20:22:54	129	1.609	03h55m (01h43m)	S. America, Europe, Africa, Asia, Australia
2019 Jan. 21	05:13:27	134	1.195	03h17m (01h02m)	Central Pacific, Americas, Europe, Africa
2021 May 26	11:19:53	121	1.009	03h07m (00h15m)	Eastern Asia, Australia, Pacific, Americas
2022 May 16	04:12:42	131	1.414	03h27m (01h25m)	Americas, Europe, Africa
2022 Nov. 08	11:00:22	136	1.359	03h40m (01h25m)	Asia, Australia, Pacific, Americas
2025 March 14	06:59:56	123	1.178	03h38m (01h05m)	Pacific, Americas, Western Europe, Western Africa
2025 Sept. 07	18:12:58	128	1.362	03h29m (01h22m)	Europe, Africa, Asia, Australia
2026 March 03	11:34:52	133	1.151	03h27m (00h58m)	Eastern Asia, Australia, Pacific, Americas
2028 Dec. 31	16:53:15	125	1.246	03h29m (01h11m)	Europe, Africa, Asia, Australia, Pacific
2029 June 26	03:23:22	130	1.844	03h40m (01h42m)	Americas, Europe, Africa, Middle East
2029 Dec. 20	22:43:12	135	1.117	03h33m (00h54m)	Americas, Europe, Africa, Asia

Umbral Magnitude is how deep into the Earth's shadow the Moon will pass (eclipse predictions by Fred Espenak, NASA's GSFC).

Beyond 2030, I run into the conundrum that all eclipse-chasers eventually face: How many more will life allow me to see?

[An astronomer must] predict [the planets'] motion in all future time, compute their orbits, determine what changes of form and position these orbits will undergo through thousands of ages, and make maps showing exactly over what cities and towns . . . an eclipse of the sun will pass fifty years hence, or over what regions it did pass thousands of years ago. A more hopeless problem than this could not be presented to the ordinary human intellect.

—ASTRONOMER SIMON NEWCOMB, 1903

CHAPTER 8

The Last Total Eclipse

Our Moon is unique. Of all the planets, only the Earth has a moon just the right size and distance to barely cover the Sun. If the Moon were smaller, or farther away, all eclipses would be annular. In such a world, the solar corona and prominences would remain forever invisible. If the Moon were larger, or closer, the Sun would still be entirely blocked, but so, too, would the corona, and once more the phenomena that have fueled the curiosity of astronomers, astrologers, and philosophers throughout history would go unseen. In either case, the existence of helium and the process that fuels the stars would have eventually been discovered through some other means, but it is hard to imagine astronomers asking questions about phenomena we would never have seen.

No other planet has such a large moon in comparison to its own size. While both Jupiter and Saturn possess moons just as big, both of those planets are also hundreds of times more massive than the Earth. Venus, our planetary twin in composition and size, has no moon at all, while Mars has nothing but the two tiny rocks of Phobos and Deimos. Only Pluto and its largest moon, Charon, are comparable double-planets to our own (never mind that Pluto is no longer a planet), but out at a distance of fifty Astronomical Units from the Sun, Charon utterly dwarfs the Sun in apparent size. It all seems so improbable.

The story of how we came to have our Moon is interwoven with the history of eclipses. While our myths and legends are filled with lunar deities that delight us with their tales of its origins, influences, and reasons for going through phases, our scientific understanding of how it came to be begins with the story of where it is now. It's a question of orbits.

Edmond Halley, whose story has been so linked to that of eclipses, was a good friend of Newton's, whose gravitational studies he helped pay to have published. Halley used these new laws of Newton's to calculate the orbits of everything from comets to the Moon, but to do so he needed a record of where they appeared in the sky and when.* For the Moon, the best data comes from the times and locations of solar eclipses, since at that moment the Moon's location is exactly known in relation to the Sun and the observer. String enough of these together, and the Moon's orbit should become obvious. This is why Halley was so

---

* This is how he found that observations of what appeared to be multiple comets appearing over a period of centuries was in fact a single comet returning every seventy-six years. Now known as Halley's Comet, it is the most famous comet in the world.

eager to have the public see and record the duration of the 1715 total eclipse over London.

But Halley also pored through historical records looking for other eclipses he could use to refine his calculations even further. A result of such a precise lunar orbit would be the accurate dating of eclipses in millennia-old Greek texts, pinpointing the exact dates for the great battles and events that had shaped civilization. Francis Baily, who would attempt to date the eclipse predicted by Thales in the *Histories* of Herodotus (and become famous for his own eclipse observations), concluded that the precision with which the unmistakable spectacle of a total solar eclipse could be calculated meant that "all attempts at imposition or deceit are easily detected by our knowledge of astronomy: and the unintentional errors of the historian are soon rectified and adjusted."

Strangely, the farther back in time Halley and his contemporaries looked, the worse their predictions matched the historical records. Either the records were wrong about where the eclipses were seen, or the times were wrong for when they occurred. Neither ever quite fit. Halley makes reference to this problem in a 1695 paper on the wonders of the ancient city of Palmyra in modern war-torn Syria. On the very last page of his paper, he asks if his readers would be so good as to please determine the exact positions of Palmyra, Baghdad, Cairo, and other cities where ancient eclipses had been seen, so that he could clear up some issues he was having with his calculations.

A hypothesis is only as good as the observations that constrain it. It's at the heart of the scientific method that the late nineteenth-century American astronomer Simon Newcomb was a principal force in promoting. He was a public advocate for the power that science, through a rigorous method of observation, hypothesis, and experimentation, could bring to bear on the mysteries of both space and time. He wrote, "The real test of progress is found in our constantly increased ability to foresee either the course of nature or the effects of any accidental or artificial combination of causes. So long as prediction is not possible, the desires of the investigator remain unsatisfied. When certainty of prediction is once attained, and the laws on which the prediction is founded are stated in their simplest form, the work of science is complete."

But the future can't be predicted if the past can't be explained; it's the heart of what a successful scientific theory must do. Newcomb felt that unless Newton's laws could accurately date ancient eclipses, it would be impossible to accurately predict future ones and thus understand how the Moon behaved. Unfortunately, the quality of eclipse records hundreds and thousands of years in the past was not always of the quality necessary to pin down exactly where the Moon had eclipsed the Sun and by how much and at what time. Trying to tease the necessary astronomical information from dusty parchment scrolls written by scribes reporting on events possibly second- or third-hand was tremendously tricky work. According to Newcomb, "There are tens of thousands of men who could be successful in all the ordinary walks of life, hundreds who could wield empires, thousands who could gain wealth, for one who could take up this astronomical problem with any hope of success. The men who have done it are therefore in intellect the select few of the human race, an aristocracy ranking above all others in the scale of being."

Newcomb was not a modest man, though he was largely self-taught from the backwoods of Nova Scotia. But he was unflinchingly dedicated to the power of physics, provided one was absolutely precise in one's observations. During the mania over the putative planet Vulcan, he was a meticulous observer, and when he failed to detect any such planet, he was critical of others' claims of discovery. Even Halley's "citizen-science" timings of the 1715 eclipse fell short in Newcomb's strictly mathematical eye. "Halley, notwithstanding his scientific merits in some directions," he wrote, "seems to have been extremely unskilled in every branch in the art of practical astronomy." So Newcomb sifted the historical records for solar and lunar eclipses recorded with precision, going so far as to travel to the Paris Observatory in the midst of the Franco-Prussian War to recover the original observation logs for eclipses that had occurred during the previous two hundred years. With bombs bursting in the distance, he confirmed what earlier astronomers as far back as Halley had suspected: the Moon is accelerating. His new calculations derived from the best observations yielded the most precise value yet for just how much. And just as important, he realized that part of the problem was that the Earth itself was slowing down, so much so that with every century, our days grow two milliseconds longer. This is really what Halley had discovered when the time of day he calculated for ancient eclipses didn't quite match with when or where they had been seen.

Lunar tides are why this is happening. Since gravity grows weaker with distance, the Moon attracts the Earth by different amounts from one side of our planet to the other. Rock doesn't deform much—at least not under the Moon's feeble mass—but water is fluid. The Moon therefore raises two liquid bulges on the Earth, one on the near side—as the oceans are pulled up from the planet—a second on the far side, as the planet is pulled out from underneath the sea. As the Earth spins on its axis, shorelines encountering these bulges experience two high and low tides every day. For any person standing on the seashore, the evidence for astronomy is present in the perpetual flow of the tides.

Since the Earth turns faster than the Moon orbits, friction with the seafloor drags the tidal bulges along with it. The greater the friction between seafloor and sea, the more energy the Earth loses and the slower it turns. The continental shelves, like the shallow Bering Strait, where North America and Asia once connected, are places of particular friction; they are the continent-sized speed bumps that the tidal bulges keep hitting. The effect, however, is not solely on the Earth. As the seafloor drags the bulge along with it, its gravity leads the Moon, pulling it forward. This constant acceleration sends the Moon spiraling away. Since the two effects are tied together, the rate at which the Earth slows down mirrors the rate at which the Moon grows more distant. Reflectors placed on the Moon by the Apollo astronauts confirm that today we are losing the Moon at a rate of 1.5 inches per year, about the same rate as fingernails grow. While that may not seem like much, it's the first piece of evidence to suggest how the Moon once formed.

George H. Darwin, son of Charles Darwin, was the first to mathematically work out the evolution of these tides. In 1878, based on the Moon's rate of recession, he found that it was only 60 million years ago that the Earth and the Moon must have been in contact—with the Moon orbiting a rapidly spinning Earth, and each taking only six hours to complete a full turn. To Darwin, the implication was clear: the Earth and Moon were once a single, rapidly spinning body that split into two when a chunk of the crust was flung into space. As evidence, he wrote of the tidal forces from the newly formed Moon, which he said would have raised the great ridges of liquid rock on the Earth that we now see as continents. The Pacific Ocean was the basin left over when the Moon split away.

The sciences of radiometric dating and continental drift wouldn't be proposed for another three decades, so Darwin had no way of knowing the Earth was much older and the continents much younger than his physics suggested. In addition, later physicists could find no plausible mechanism that could get the Earth to spin so fast that it would break into two. Still, even as late as the 1970s, the fission hypothesis found some supporters, including the American chemist Harold Urey.

Urey, a Nobel Laureate at the University of Chicago, was one of the first to propose that the Apollo moon rocks could be used to understand how the Moon and solar system formed. By understanding the mechanism by which planetary bodies formed, he hoped to be able to say how rare they might be in the cosmos. By the end of the Apollo missions, three principal findings had come out of the rocks the astronauts brought back. One was that lunar rocks have low amounts of volatiles, elements that can be easily vaporized or boiled away. Second, lunar rocks are depleted in metals. On average, the Moon has much less metal than the Earth and a metallic core that, if it exists, is smaller in relation to its size than the Earth's. Both of these conclusions speak to a world somewhat different from our own. The third finding was that the ratio of certain elemental isotopes—oxygen, in particular—was almost identical on the Earth and the Moon.* Since oxygen isotope ratios in Earth rocks are wildly different from those in meteorites from the asteroid belt or from Mars, this similarity tells us that the Earth and the Moon must share a common event in their origin.

By the end of the Apollo missions, no single hypothesis had yet been proposed to explain all of these findings. Some

---

* Different elements have different numbers of protons in their nucleus. Different isotopes of a single element have the same number of protons, but different numbers of neutrons.

scientists thought the Earth and the Moon formed together, while others thought the Earth must have captured the Moon from elsewhere; some, like Urey, still favored a Moon that formed out of the Earth. Into this confusion of hypotheses, the astronomers Alastair Cameron and William Ward proposed a revolutionary idea in 1976: What if the Moon really had split off from the Earth, but for a reason that had nothing whatsoever to do with the Earth's rotation? Cameron was a Canadian astrophysicist who had written extensively on the origin and evolution of planets and stars. He and Ward proposed that, yes, the Moon had spun away from the Earth, but only after the Earth was hit by another planet. The Moon is the result of the violence of worlds.

As much as their idea may seem the stuff of science fiction, it does follow naturally from how astronomers think planets should form. Stars and planets coalesce out of clouds of spinning gas and dust. Over time, dust in these cosmic clouds is drawn together by gravity. Over millions of years of matter colliding and sticking, small things grow to become big things. Within such a hypothesis, the last stage of such collisions would surely have seen a few large impacts from nearly planet-sized objects, called *planetesimals*. The impact that formed the Moon was simply one of the last our planet experienced.

Imagine the Earth four and a half billion years ago as a hot, rapidly spinning world of glowing rock. Inside, the majority of high-density metals had already settled into a molten core, leaving lower-density rock floating on top to form a mantle and a crust. On such a hot world, volcanoes would have spewed lava across a barren surface, their plumes releasing the first gases that would form an atmosphere. Someone standing on such a hellish world, looking high overhead, would have seen a star-filled expanse in which no moon yet existed. Then one night, one of those millions of stars was noticeably brighter than it had been the night before (although the time between sunsets was no more than five hours). Soon, had anyone been there to see it, the star would have become a disk the same size as Mars. Planetary scientists have nicknamed this world Theia after the mother of Selene, the goddess of the Moon in Greek mythology.

As the two worlds approached each other, the tidal forces between them would have grown so much that as our planet bulged outward, earthquakes would have rocked the landscape. On the fatal final day, when the two planets struck in a glancing blow, one entire hemisphere of Theia would have instantly vaporized, exploding outward and carrying massive amounts of molten debris into orbit around the now rapidly coalescing remnants of the two previous worlds. The new planet that was created on that day is the Earth on which we all now live, and for a brief period it possessed a ring out of which our Moon eventually formed.

This is a hypothesis that explains all that has been seen. From a glancing blow, computer models reveal that the Moon would have formed out of the relatively low-density elements in the mantles of the two planets. The violence of the collision would have vaporized all the volatiles in the newly created Moon, including most of its water. Since the two new worlds formed from the same hot molten mix of planets, the Earth and the Moon now exhibit nearly identical isotope ratios.

Had that been the last major collision our Earth ever encountered, our Moon would now orbit in the same plane as the Earth around the Sun, and we'd experience solar eclipses every month. But recent computer models reveal that over the next few tens of millions of years, any remaining solar system debris nearly as large as the Moon that passed too close to our double-world would have contributed just enough of a gravitational tug to send the Moon's orbit tilting at the angle at which we now find it. The evidence for these close flybys and occasional impacts is found in our precious metals. Precious metals like platinum and gold react strongly with iron. On a molten Earth, high-density iron would sink to the core and take these elements with them. But the collision of even half a dozen lunar-sized objects following the formation of our Moon would have brought just enough additional gold and other precious metals to litter our surface in the amounts we now see. In the words of the planetary scientist Robin Canup, "Had such a population of objects not existed, the Moon might be orbiting in Earth's orbital plane, with total solar eclipses occurring as a spectacular monthly event. But our jewelry would be much less impressive—made from tin and copper, rather than from platinum and gold." Eclipses are literally worth a world of wealth.

From the moment of the Moon's creation, as the bodies cooled and tidal friction grew, the rapidly spinning Earth slowed and the Moon began its long spiral away. We will never lose the Moon. But a day will come when the Earth has slowed so much that it turns at the same rate as the Moon orbits, and a day on Earth will be as long as a month. The Earth will be tidally locked to the Moon as the Moon is now locked to the Earth, and a single hemisphere on each world will forever stare at the other. When that happens, both the Moon and the Earth as seen from the other will never move in the sky, never rising or setting.

By that time, the Moon will be so far away that it no longer fully covers the Sun—the last total solar eclipse will have occurred. At the rate the Moon is currently receding, we only have another 563 million years of total solar eclipses left to us. That last eclipse will be a brief one, the Moon's disk only barely the size of the Sun that it's covering. Perhaps totality will be no more than a second in length, and every one after that, for the rest of the Earth's life, will be no more than an annular eclipse. The spectacle will be gone. The total solar eclipses we now see—that were once an omen of terror, that then became a scientific tool, and that have now turned into a tourist attraction—will have been just a transitory phase in the life of our planet.

Ironically, for a phenomenon that we can talk about with such great certainty, where coronaphiles can make plans to see one decades in advance, it's impossible to say the exact date and location of that last ever eclipse. The reason is that over its lifespan the rate at which the Moon recedes has varied with time. At its current, precisely known rate, only 1.5 billion years ago the Moon should have been so close that the Earth's mantle would have melted and the Moon ripped apart. Needless to say, there is no record of this happening. The only solution is that the Moon must have been receding at a slower rate in the past. Evidence from ancient seafloor sediments shows that the length of the day has similarly fluctuated over billions of years.

All of this is due to the slow motions of continents and seafloor altering the rate at which friction slows down the Earth and speeds up the Moon. As it has changed in the past, so, too, will it change in the future, and the exact rate of recession becomes impossible to predict. But thanks to the long chain of human beings who have watched the Moon and the Sun and divined the patterns in their eclipses—including Halley and Newcomb, who used the timing of eclipses to discern the subtle motions of the Earth and the Moon—the science is clear that such a day will eventually arrive. Enjoy an eclipse while you can.

Newcomb once wrote, in relation to the predictive power of science, that "whenever the subject becomes so well understood that the chain of natural causes can be clearly followed, miracles and final causes cease, so far as the scientific explanation of things is concerned." It is ironic, then, that in an echo of the "God of the Gaps," by which we once attributed

eclipses to demons and gods, there are those who still attri-
bute eclipses to a deity. They see proof of a world designed by
a Creator in the size and distance of our Moon, and in the fact
that eclipses will have occurred on this planet only during a
time when an intelligent species was here to behold it.

Some proponents of this idea go so far as to claim that
proof of its truth will be found by looking for intelligent life
only where there are moons large enough to cause eclipses;
wherever one finds eclipses, according to this theory, one
will certainly find intelligent life.* I suppose this bodes well
for NASA's upcoming mission to Jupiter's moon Europa, as
all four of its Galilean satellites are at the right size and distances to eclipse one another every six years.

Interestingly, given the importance some creationists
place on its existence, our Moon may actually play a crucial
role in the answer to the question of how rare life may be.
Were it not for our Moon, then 4 billion years ago there
would have been few tide pools to harbor the "primordial
soup" of complex organic molecules that chemists like Urey
and Carl Sagan studied as the building blocks of life. Once
formed, these organics needed millions, then billions, of
years to make the transition from soup to sentience. That is
a journey that requires a hospitable planet, constant in its
climate over great spans of time.

Tidal forces are the reason our climate has remained relatively constant over billions of years. Mars has no similar-
sized moon, and recent evidence shows that its *obliquity*, the

---

* Google "recession of the Moon" and seven of the first ten results to come up
  are from creationist websites.

angle its pole tilts relative to its orbit, changes chaotically with time. This is important because a planet's obliquity determines the severity of its seasons. Right now, Mars's obliquity is nearly identical to Earth's, but models show that over millions of years Mars's tilt has wandered from as little as 0 degrees to as high as 80 degrees. These changes have made Mars's summer seasons so warm that water could flow across its polar surface, while creating winters so fierce that glaciers buried the surface beneath miles of ice. This changing obliquity is due to the gravitational tug of the rest of the solar system, primarily the Sun and Jupiter. We are affected by them as well, but our Moon's tidal embrace keeps our planet in check. It's hard for the tidal bulge around our equator to wobble too much if the Moon is always there to pull it back into place. The result is that the obliquity of our planet, first likely set in the collision that formed the Moon, hasn't changed by more than a degree over the past half million years. Even so, these tiny shifts have helped give rise to a cycle of gradual ice ages, the most recent of which ended only 10,000 years ago. As great as these ice ages may seem, imagine what would have happened had our poles tilted even more.* Intelligent life on Earth might only be possible in the presence of a moon like our own.

This is a sobering thought, given what such a chance occurrence our Moon may have been. What does it mean

---

* Changes in our obliquity, orbital eccentricity, and precession of the equinoxes are all part of what is known as the Milankovitch cycle of climactic change. They lead to only slow changes in climate over thousands of years, not the changes the Earth is experiencing now over the past two hundred years.

that we may only be here because of an accident? People criticize science for making us small, for taking us from the center of creation and demoting us to tiny inhabitants in space and time of a vastly larger impersonal universe.

Yet human history is inextricably linked with our need to accurately understand the world in which we live. From the time of our earliest ancestors, as they hunted for food and then developed agriculture, we would not have survived to the present without having unlocked the mysteries of the seasons and the clockwork patterns of the sky. This is a process that need not be opposed to religion, as astronomers going back to Kepler (a Protestant) and Galileo (a Catholic) attest. The current director of the Vatican Observatory, Brother Guy Consolmagno, a Catholic Brother and PhD planetary scientist, once told me that the proof of God is in the fact that the universe is knowable.

So what do we know? We know that we occupy a special place, on a special planet, with a special moon, during a special moment in history. We are privileged to inhabit a habitable world that has been capable of sustaining complex life over billions of years, thanks to a moon created by a chance event. But that chance event is a natural part of how solar systems form—and we now know of over a thousand others out there in space, with more being discovered every year.

All this we know, in part, due to eclipses. Through these momentary alignments, the inhabitants of this planet— in ancient Chinese palaces, Babylonian observatories,

Mediterranean cities, and Mayan temples—all deduced that our world is one of repeating patterns that are knowable. From these beginnings we have followed the path of totality to every corner of the Earth to learn more about the Moon, the Sun, and a galaxy of planets. The secrets of the universe have been revealed by shadows stretching over the light-years between stars. Because of these shadows, we now have a galactic context in which to understand how common planets may be, while realizing, as we look around at our own solar system, how utterly inhospitable most planets are. While we may not be alone in the universe, we are still precious, as is the world that sustains us. All this we see when the Sun disappears behind the Moon and for a brief moment each one of us is aligned with the heavens.

Acknowledgments

A work like this would not be possible without the advice, assistance, and contributions of many experts across many fields. For their generous support and effort I'd like to thank Amy Sayle at Morehead Planetarium, who read every word of this book two or three times; Ed Krupp at Griffith Observatory; psychologist and coronaphile Kate Russo; eclipse-chaser and photographer Alan Dyer; Barbara Cohen at NASA Marshall Space Flight Center; archeoastronomer Kim Malville; Eanna Flanagan at Cornell University; Devon Belcher at Oglethorpe University; Craig DeForest at Southwest Research Institute; and William Abernathy in San Francisco. For photographs, maps, and invaluable assistance I'd like to thank eclipse photographer Geoff Sims; filmmaker David Makepeace; Angela Speck at the University of Missouri; Doug Duncan at the Fiske Planetarium; Paul B. Jones and the Staunton River Star Party; Amy Barr at the Planetary Science Institute; Michael Zeiler, mapmaker extraordinaire at GreatAmericanEclipse.com; John Tilley for his worldwide eclipse location calculations; Fred Espenak, who is Mr. Eclipse; Jay Pasachoff and his invaluable comments; Lis Mortensen of the National Museum of the Faroe Islands; and my mom, Kareen Tierney, and Mike Tierney, who scouted out eclipse locations in eastern Oregon near their home in Condon. I'd also like to thank my crewmates on *Star Flyer* for the total eclipse of 2013 (my apologies to everyone I interviewed on board for this book—I lost my iPhone rafting the Grand Canyon before I could transcribe them); National Park Ranger Sonya Popelka, who suggested the book's title; and Lua Gregory of the University of Redlands Library, without whom I would not have half my sources and images. Lastly, thank you to my publishers, TJ Kelleher and Quynh Do, who greatly improved the text and had confidence in the project; and my agent, Farley Chase, who suggested the idea for this book, and without whom it would not exist. Any errors in this work are through no fault of theirs. To my wife, Julie Rathbun, it is my greatest hope that in 2017 you will soon see what all the trouble was worth.

Solar Eclipse Resources

General Eclipse Information

The official NASA website for information about the 2017 total solar eclipse: http://eclipse2017.nasa.gov

NASA's Goddard Space Flight Center Solar Eclipse Resource: http://eclipse.gsfc.nasa.gov/solar.html

Eclipse Travel Posters for 2017 (and beyond): www.tylernordgren.com/worldwide-solar-eclipse

Fred Espenak's General Informaton and How-To Site: MrEclipse.com

Eclipse Maps

Eclipse-Maps.com: www.eclipse-maps.com

Michael Zeiler's 2017 eclipse website: www.GreatAmericanEclipse.com

Interactive Google Maps by Xavier M. Jubier: http://xjubier.free.fr/en/site_pages/SolarEclipsesGoogleMaps.html

Eclipse Glasses and Solar Filters

NASA Eclipse Safety: http://eclipse.gsfc.nasa.gov/SEhelp/safety2.html

Rainbow Symphony Solar Eclipse Glasses: www.rainbowsymphonystore.com/eclipseshades.html

Thousand Oaks Optical: http://thousandoaksoptical.com/index.html

Eclipse Photography

NASA's Goddard Space Flight Center Eclipse Photography Page: http://eclipse.gsfc.nasa.gov/SEhelp/SEphoto.html

Eclipse2017 Photography Website: www.eclipse2017.org/2017 /photographing.htm

Dr. Jay Pasachoff's Eclipse Links: http://totalsolareclipse.org

Eclipse Books

Leon Golub and Jay M. Pasachoff, *Nearest Star: The Exciting Science of Our Sun*, 2nd ed. New York: Cambridge University Press, 2014.

Jay M. Pasachoff, *Peterson Field Guide to the Stars and Planets*, 4th ed. New York: Houghton Mifflin Harcourt, 2016.

Dennis Schatz and Andrew Fraknoi, *Solar Science—Exploring Sunspots, Seasons, Eclipses, and More*. Arlington, VA: National Science Teachers Association, NSTA Press, 2015.

Notes

Chapter 1: A Day with Two Dawns and Midnight at Noon

9 **Ancient Chinese eclipse accounts:** E. C. Krupp, *Beyond the Blue Horizon: Myths and Legends of the Sun, Moon, Stars, and Planets* (New York: Oxford University Press, 1991), 158–162.

10 **The French astronomer and historian:** J.-P. Verdet, *The Sky: Mystery, Magic, and Myth*, trans. Anthony Zielonka (New York: Harry N. Abrams, 1992), 73.

10 **In India, the people banged pots:** M. Littman, F. Espenak, and K. Wilcox, *Totality: Eclipses of the Sun* (New York: Oxford University Press, 2009), 40.

10 **"common folk raised a cry . . . ":** Krupp, *Beyond the Blue Horizon*, 162.

11 **paths slowly spiral:** J. Meeus, *Mathematical Astronomical Morsels* (Richmond: Willmann-Bell, 1997), 88–92.

12 **one of *three* solar eclipses:** W. B. Masse and R. Soklow, "Black Suns and Dark Times: Cultural Responses to Solar Eclipses in the Ancient Puebloan Southwest," in *Current Studies in Archaeoastronomy: Conversations Across Space and Time*, eds. J. W. Fountain and R. M. Sinclair (Durham, NC: Carolina Academic Press, 2005), 57.

12 **To the upper left:** J. M. Vaquero and J. M. Malville, "On the Solar Petroglyph in the Chaco Canyon," *Mediterranean Archaeology and Archaeometry* 14, no. 3 (2014): 189–196.

13 **monumental "Great Houses":** Ibid., 196.

13 **"The air became very still . . . ":** W. Keeler, "Sharp Rays: Javanese Responses to a Solar Eclipse," *Indonesia* 46 (1998): 46, 91.

14 **dared not even look:** Ibid., 91.

15 **"[If the king] does not share . . . ":** P. K. Wang and G. L. Siscoe, "Ancient Chinese Observations of Physical Phenomena Attending Solar Eclipses," *Solar Physics* 66 (1980): 190–191.

16 **adding an eclipse:** D. Henige, "Day Was of Sudden Turned into Night: On the Use of Eclipses for Dating Oral History," *Comparative Studies in Society and History* 18, no. 4 (1976): 489–491.

17 **"the doctor could predict . . . ":** D. C. Poole, *Among the Sioux of Dakota* (New York: D. Van Ostrand, 1881), 76–77.

21 **"cognitive fluidity":** L. H. Robbins, "Astronomy and Prehistory," in *Astronomy Across Cultures: The History of Non-Western Astronomy*, ed. H. Selin (Great Britain: Kluwer Academic, 2000), 37.

21 **Evidence for this fluidity:** Ibid., 37.

21 **Less than a mile from the Nile River:** Ibid., 38–39.

Chapter 2: Two Worlds One Sun

34 **Moon's 29.5-day lunation:** M. Littman, F. Espenak, and K. Willcox, *Totality: Eclipses of the Sun* (New York: Oxford University Press, 2009), 29–33.

34 **alignment of the Great Houses:** J. M. Malville and C. Putnam, *Prehistoric Astronomy in the Southwest* (Boulder: Johnson Books, 1993), 27–38.

36 *eclipse season:* F. Espenak, "NASA Eclipse Web Site," Goddard Space Flight Center, http://eclipse.gsfc.nasa.gov/SEsaros/SEperiodicity.html.

36 **great fight in the sky:** Littman et al., *Totality*, 268–269.

37 **With little more than these measures:** F. Hoyle, *On Stonehenge* (London: Heineman, 1977), 43–52.

37 **Dresden Codex:** E. C. Krupp, *Beyond the Blue Horizon: Myths and Legends of the Sun, Moon, Stars, and Planets* (New York: Oxford University Press, 1991), 162.

39 **35 more before it is done:** Littman et al., *Totality*, 268–269.

39–40 **Their observations were so precise:** M. Ossendrijver, "Ancient Babylonian Astronomers Calculated Jupiter's Position from the Area Under a Time-Velocity Graph," *Science* 351, no. 6272 (2016): 482–484.

40 **As far back as the seventh century:** F. Rochberg-Halton, "Review Articles: The Babylonian Astronomical Diaries," *Journal of the American Oriental Society* 111, no. 2 (1991): 324–326.

40 **perfect agreement:** A. Aaboe, J. P. Britton, J. A. Henderson, O. Neugebauer, and A. J. Sachs, "Saros Cycle Dates and Related Babylonian Astronomical Texts," *Transactions of the American Philosophical Society*, new ser. 81, no. 6 (1991): 1–31.

41 **"In this war they brought . . . ":** Herodotus, *The Histories*, trans. Thomas Worthen, Department of Classics, University of Arizona, 1997, https://scholar.lib.vt.edu/ejournals/ElAnt/V3N7/worthen.html.

41 **"day turned to night":** F. Baily, "On the Solar Eclipse Which Is Said to Have Been Predicted by Thales," *Philosophical Transactions of the Royal Society of London* 1 (1811): 400–401.

42 **World Astronomical Heritage Site:** M. E. Ozel, "A Unique Astronomical Heritage Place: The 28 May 585 BCE Solar Eclipse," in *The Role of Astronomy in Society and Culture, Proceedings of IAU Symposium No. 260*, eds. D. Valls-Gabaud and A. Boksenberg, International Astronomical Union, 2009.

42 **whether Thales actually predicted:** D. Panchenko, "Thale's Prediction of a Solar Eclipse," *Journal for the History of Astronomy* 25 (1994): 275–277.

42 **Like their colleagues in Babylon:** D. Henige, "Day Was of Sudden Turned into Night: On the Use of Eclipses for Dating Oral History," *Comparative Studies in Society and History* 18, no. 4 (1976): 484.

45 **It works, but:** National Science Foundation, "Correct Answers to Science Literacy Questions, Science and Engineering Indicators," 2006, www.nsf.gov/statistics/seind06/append/c7/ato7–10.pdf.

47 **"The seeker after the truth . . . ":** A. I. Sabra, "Ibn al-Haytham: Brief Life of an Arab Mathematician: Died Circa

1040," *Harvard Magazine*, 2003, based on Ibn al-Haytham's *Book of Optics and Aporias Against Ptolemy*, trans. A. I. Sabra.

49 **refuted the prevailing claim:** M. A. Smith, "What Is the History of Medieval Optics Really About," *Proceedings of the American Physical Society* 148, no. 2 (2004): 182.

49 **We use this technique today:** A. I. Sabra, *Ibn al-Haytham*, *Dictionary of Scientific Biography*, vol. 6, eds. Jean Hachette and Joseph Hyrtl (New York: Charles Scribner's and Sons, 1972), 190.

50 **"The undoubted truth . . . ":** A. I. Sabra, "Configuring the Universe: Aporetic, Problem Solving, and Kinematic Modeling as Themes of Arabic Astronomy," *Perspectives on Science* 6, no. 3 (1998): 288.

Chapter 3: Shadows Across a Sea of Stars

59 **The locations of all those:** D. W. Graham and E. Hintz, "Anaxagoras and the Solar Eclipse of 478 BC," *Apeiron: A Journal for Ancient Philosophy and Science* 40, no. 4 (2007): 321–335.

59 **"Anaxagoras [says that the Moon] is . . . ":** Plutarch, *De fac orb lun*, 932a.

59 **"the sun exceeds . . . ":** Hippolytus, *Refutation of All Heresies*, I.8.8.

60 *bematists:* I. Fischer, "Another Look at Eratosthenes' and Posidonius' Determinations of the Earth's Circumference," *Quarterly Journal of the Royal Astronomical Society* 16 (1975): 152–153.

62 **by as little as 2 percent:** Ibid., 153–160.

63 **Antikythera Device:** C. C. Carman and J. Evans, "On the Epoch of the Antikythera Mechanism and Its Eclipse Predictor," *Archive for History of Exact Sciences* 68 (2014): 697–700.

64 **1,500 years before metal gears:** Ibid., 763.

64 **it wasn't Columbus's beliefs:** Fischer, "Another Look," 164.

65 **The actual distance:** V. F. Rickey, "How Columbus Encountered America," *Mathematics Magazine* 65 (1992): 223–224.

65 **Columbus thought:** Ibid., 223–224.

65 **In reality, Columbus was only:** D. W. Olson, "Columbus and an Eclipse of the Moon," *Sky & Telescope*, October 1992, 437–438.

65 **a lucky combination:** Ibid., 437–440.

66 **Meriwether Lewis and William Clark:** R. S. Preston, "The Accuracy of the Astronomical Observations of Lewis and Clark," *Proceedings of the American Philosophical Society* 144, no. 2 (2000): 176.

66 **"Observed an Eclips . . . ":** M. Lewis, W. Clark, and Members of the Corps of Discovery, "The Journals of the Lewis and Clark Expedition: January 14, 1805," ed. G. Moulton (Lincoln: University of Nebraska Press, 2002), http://lewisandclarkjournals .unl.edu/read/?_xmlsrc=1805–01–14.xml&_xslsrc=LCstyles.xsl.

66 **astronomical almanacs:** A. J. Large, "How Far West Am I? The Almanac as an Explorer's Yardstick," *Great Plains Quarterly* 13, no. 2 (1993): 119.

67 **"Major & brother . . . ":** M. P. Ghiglieri, *First Through Grand Canyon: The Secret Journals and Letters of the 1869 Crew Who Explored the Green and Colorado Rivers* (Flagstaff, AZ: Puma Press, 2010), 195.

68 **"complained that he was losing . . . ":** D. Sobel, *Longitude: The True Story of a Lone Genius Who Solved the Greatest Scientific Problem of His Time* (New York: Walker and Company, 2007), 27.

68 **location along the Pacific Coast:** Large, "How Far West Am I?," 121.

68 **position of the Rocky Mountains:** Ibid., 122.

69 **Aristarchus measured:** J. L. Berggren and N. Sidoli, "Aristarchus's On the Sizes and Distances of the Sun and Moon: Greek and Arabic Texts," *Archive for History of Exact Sciences* 61, no. 3 (2007): 213–254.

73 **"I would have several . . . ":** E. Halley, "Methodus singularis quâ Solis Parallaxis sive distantia à Terra, ope Veneris intra Solem conspiciendae, tuto determinare poterit," *Philosophical Transactions of the Royal Society* 29 (1716): 454. English translation: "A New Method of Determining the Parallax of the

Sun, or His Distance from the Earth," *Abridged Transactions of the Royal Society* 6 (1809): 246.

74 **"Saturday, 3rd [June 1769] . . . ":** J. Cook, *First Voyage Round the World: Captain Cook's Journal During His First Voyage Round the World Made in H.M.* Bark (Barsinghausen, Germany: Unikum Verlag, 2011), 139.

76 **"is probably the longest . . . ":** H. S. Hogg, "Out of Old Books: Le Gentil and the Transits of Venus 1761 and 1769," *Journal of the Royal Astronomical Society of Canada* 45 (1951): 37.

76 **transit expeditions:** S. J. Dick, *Sky and Ocean Joined: The U.S. Naval Observatory 1830–2000* (Cambridge, UK: Cambridge University Press, 2000), 243.

76 **92,885,000 miles:** Ibid., 241.

77 **two places at once:** M. Romer and I. B. Cohen, "Roemer and the First Determination of the Velocity of Light (1676)," *Isis* 31, no. 2 (1940): 329.

77 **thing was impossible:** Ibid., 334–335.

78 **11 minutes per AU:** O. Rømer, "A Demonstration Concerning Motion of Light, Communicated from Paris, in the *Journal des Scavans,* and Here Made in English," *Philosophical Transactions of the Royal Society* 12 (1677): 893–894.

78–79 **658,000 AU away:** F. W. Bessel, "On the Parallax of 61 Cygni," *Monthly Notices of the Royal Astronomical Society* 4 (1838): 160.

80 **"Earth 2.0":** J. M. Jenkins et al., "Discovery and Validation of Kepler-452b: A 1.6-R⊕ Super Earth Exoplanet in the Habitable Zone of a G2 Star," *Astronomical Journal* 150, no. 2 (2015): 56.

81 **sniffing alien air:** D. Charbonneau, T. M. Brown, R. W. Noyes, and R. L. Gilliland, "Detection of an Extrasolar Planet Atmosphere," *Astrophysical Journal* 568 (2001): 377.

81 **chemical signature:** L. Arnold et al., "The Earth as an Extrasolar Transiting Planet. II. HARPS and UVES Detection of Water Vapour, Biogenic O_2, and O_3," *Astronomy and Astrophysics* 564 (2014): 58.

Chapter 4: As Below, So Above

86 **wasn't solid at all:** S. Y. Edgerton, "Galileo, Florentine 'Disegno,' and the 'Strange Spottedness' of the Moon," *Art*

Journal: Art and Science: Part II, Physical Sciences 44, no. 3 (1984): 226.

87 **Galileo, raised in the heart:** Edgerton, "Galileo, Florentine," 225–227.

87 **"[I] have been led . . . ":** G. Galilei, Sidereus Nuncius; or the Sidereal Messenger, trans. A. Van Helden (Chicago: University of Chicago Press, 1989), 40.

87 **Galileo does not say:** Ibid., 38.

87 **most exciting moment:** T. E. Nordgren et al., "Calibration of the Barnes-Evans Relation Using Interferometric Observations of Cepheids," Astronomical Journal 123, no. 6 (2002): 3380–3386.

89 **Chinese records:** G. Galilei and C. Scheiner, On Sunspots, trans. E. Reeves and A. Van Helden (Chicago: University of Chicago Press, 2010), 9.

89 **Indian Kashi Khanda text:** J. M. Malville and R. P. B. Singh, "Visual Astronomy in the Mythology and Ritual of India: The Sun Temples of Varanasi," Vistas in Astronomy 39, no. 4 (1995): 443.

90 **His letters from this time:** F. Baily, Journal of a Tour in Unsettled Parts of North America in 1796 & 1797 (London: Baily Brothers, 1856), 75–320.

90 **accepted a position:** Ibid., 70–71.

90 **calculated the date:** F. Baily, "On the Solar Eclipse Which Is Said to Have Been Predicted by Thales," Philosophical Transactions of the Royal Society of London 1 (1811): 234.

90 **Twenty-five years later:** F. Baily, "On a Remarkable Phenomenon That Occurs in Total and Annular Eclipses of the Sun," Memoirs of the Royal Astronomical Society 10 (1838): 31.

91 **bright "beads":** Ibid., 5.

91 **He got close enough:** J. Williamson, "Total Solar Eclipse of October 1780," Bangor Historical Magazine 6 (1891): 63–65.

91 **bright points of light:** D. Steel, Eclipse: The Celestial Phenomenon That Changed the Course of History (Washington, DC: Joseph Henry Press, 2001), 177–179.

92 **"I was astounded . . . ":** F. Baily, "Some Remarks on the Total Eclipse of the Sun, on July 8th 1842," Monthly Notices of the

Royal Astronomical Society 15 (1846): 4. Unless otherwise noted, italics are reproduced from the original quotations.

93 **measured the corona's extent:** J. J. de Ferrer, "Observations of the Eclipse of the Sun, June 16th, 1806, Made at Kinderhook, in the State of New York," *Transactions of the American Philosophical Society* 6 (1806): 274.

93 **three large red "protuberances":** Baily, "Some Remarks on the Total Eclipse," 6.

95 **The entire expedition equipment list:** W. De la Rue, "The Bakerian Lecture: On the Total Solar Eclipse of July 18th, 1860, Observed at Rivabellosa, Near Miranda de Ebro, in Spain," *Philosophical Transactions of the Royal Society of London* 152 (1862): 337.

96 **"In many cases . . . ":** H. Rothermel, "Images of the Sun: Warren De la Rue, George Biddell Airy and Celestial Photograph," *British Society for the History of Science* 26, no. 2 (1993): 137.

97 **"We can imagine . . . ":** A. Comte, *The Positive Philosophy of Auguste Comte,* vol. 2, ed. J. Chapman (London: Chapman, 1853), 9.

100 **another twenty-four years:** E. B. Frost, "Helium, Astronomically Considered," *Publications of the Astronomical Society of the Pacific* 7, no. 45 (1895): 317.

101 **agreed well with his calculation:** C. Darwin, *On the Origin of Species by Means of Natural Selection, or the Preservation of Favoured Races in the Struggle for Life* (London, 1859), 282–287.

102 **Fusion was what fueled the Sun:** A. S. Eddington, "The Internal Constitution of the Stars," *Observatory* 43 (1920): 353–355.

104 **"coronium":** C. Young, "The Wave-Length of the Corona Line," *Astrophysical Journal* 10 (1899): 306.

107 **"If, indeed, the sub-atomic . . . ":** Eddington, "Internal Constitution," 355.

Chapter 5: The Eclipse That Changed the World

112 **found precisely where Le Verrier said it would be:** N. W. Hanson, "Leverrier: The Zenith and Nadir of Newtonian Mechanics," *Isis* 53, no. 3 (1962): 363.

112 **"forces that set nature in motion . . . "**: P. S. Laplace, *A Philosophical Essay on Probabilities* (New York: J. Wiley, 1902).

114 **"His telescope was a small one . . . "**: "Vulcan," *New York Times*, May 27, 1873.

115 **"searched that region thoroughly . . . "**: "Was It the Intra-Mercurial Planet?" *Astronomical Register* 7 (1869): 227–228.

116 **"with sneering astronomic smiles . . . "**: "Vulcan," *New York Times*, September 26, 1876.

116 **"I had committed to memory . . . "**: J. Rodgers, "Letters Relating to the Discovery of Intra-Mercurial Planets," *Astronomiche Nachrichten* 93 (1878): 162.

116 **"about one minute after totality . . . "**: Ibid., 164–165.

117 **"one brilliant discovery . . . "**: C. A. Young, "Vulcan and the Corona," *New York Times*, August 16, 1878.

117 **calculated Vulcan's size:** Ibid.

117 **locations didn't match:** Rodgers, "Letters Relating to the Discovery of Intra-Mercurial Planets," 163–164.

117 **"Prof. Swift arrived . . . "**: "The Recent Solar Eclipse," *New York Times*, August 4, 1878.

118 **"The great eclipse of 1886 . . . "**: L. Swift, "The Intra-Mercurial Planet Question Not Settled," *Sidereal Messenger* 2 (1883): 242–243.

118 **belt of debris:** S. Newcomb, "Note on Accounting for the Secular Variations of the Orbits of Venus and Mercury," *Astronomical Journal* 14 (1894): 117–118.

118 **simple ring of dust:** C. D. Perrine, "The Search for an Intra-Mercurial Planet at the Total Solar Eclipse of 1901, May 18," *Publications of the Astronomical Society of the Pacific* 14 (1902): 160–163.

118 **unsettling possibility:** A. Hall, "A Suggestion in the Theory of Mercury," *Astronomical Journal* 14 (1894): 49.

121 **"The more important . . . "**: A. A. Michelson, *Light Waves and Their Uses* (Chicago: University of Chicago Press, 1903).

125 **Einstein was influenced:** A. I. Miller, *Einstein, Picasso: Space, Time, and the Beauty That Causes Havoc* (New York: Basic Books, 2001), 2.

130 **instruments confiscated:** J. Earman and C. Glymour, "Relativity and Eclipses: The British Eclipse Expeditions of 1919 and Their Predecessors," *Historical Studies in the Physical Sciences* 11, no. 1 (1980): 62.

130 **impounded their instruments:** Ibid., 62–63.

130 **"No quotations from German . . . ":** Lord Walsingham, "German Naturalists and Nomenclature," *Nature* 102, no. 2549 (1918): 4.

130 **"for bold initiative . . . ":** M. G. Boccardi, "German Science, and Latin Science," *Observatory* 39 (1916): 384–385.

132 **"When the Einstein theory . . . ":** Earman and Glymour, "Relativity and Eclipses," 66.

132 **devout Quaker:** Ibid., 71.

133 **"Eddington was deferred . . . ":** S. Chandrasekhar, "Verifying the Theory of Relativity," *Bulletin of Atomic Scientists* 31 (1975): 18.

133 **England alone:** Earman and Glymour, "Relativity and Eclipses," 70–71.

135 **To prepare the public:** A. Sponsel, "Constructing a 'Revolution in Science': The Campaign to Promote a Favourable Reception for the 1919 Solar Eclipse Experiments," *British Journal for the History of Science* 35, no. 4 (2002): 443–444.

135 **"further telegrams . . . ":** Ibid., 444–445.

136 **"The result was contrary . . . ":** D. Kennefick, "Testing Relativity from the 1919 Eclipse: A Question of Bias," *Physics Today* 62 (2009): 40–41.

136 **presented the findings:** Sponsel, "Constructing a 'Revolution in Science,'" 455–459.

136 **Dyson and Eddington announced:** Ibid., 460–461.

137 **"If the results obtained . . . ":** J. J. Thompson, "Joint Eclipse Meeting of the Royal Society and Royal Astronomical Society," *Observatory* 42 (1919): 394.

137 **"REVOLUTION IN SCIENCE . . . ":** *London Times*, November 7, 1919.

137 **"LIGHTS ALL ASKEW . . . ":** *New York Times*, November 10, 1919.

138 **"decisive" experiment:** Earman and Glymour, "Relativity and Eclipses," 84.

139 **Greenwich Observatory confirmed:** G. M. Harvey, "Gravitational Deflection of Light," *Observatory* 99 (1979): 195–198.

139 **But the testing and retesting:** Ibid., 42.

139 **Precision far greater:** Ibid., 41.

139 **Astronomers continue to subject the details:** B. P. Abbott et al., "Observation of Gravitational Waves from a Binary Black Hole Merger," *Physical Review Letters* 116 (2016): 061102.

Chapter 6: Saros Siblings

148 **traveled to London:** E. Halley, "Observations of the Late Total Eclipse of the Sun on the 22d of April Last Past, Made Before the Royal Society at Their House in the Crane-Court in Fleet-Street, London. By Dr. Edmund Halley, Reg. Soc. Secr. with an Account of What Has Been Communicated from Abroad Concerning the Same," *Philosophical Transactions* (1715): 251.

152 **"THAT 'DIAMOND RING' . . . ":** "That 'Diamond' Ring in the Sun's Eclipse," *New York Times*, January 28, 1925, 18.

153 **largest formation of airplanes:** "35 Airplanes to Take a Big Scientific Party," *New York Times*, January 24, 1925, 2.

153 **"As the machines winged . . . ":** "View the Eclipse from 25 Airplanes," *New York Times*, January 25, 1925, 2.

154 **theaters on Broadway:** "New Haven Buzzes with Eclipse Talk," *New York Times*, January 24, 1925, 2.

155 **The Concorde traveled:** B. Mulkin, "In Flight: The Story of Los Alamos Eclipse Expeditions," *Los Alamos Science* (Winter/Spring 1981), 72.

158 **"before an eclipse . . . ":** A. Soojung-Kim Pang, *Empire and the Sun: Victorian Solar Eclipse Expeditions* (Stanford, CA: Stanford University Press, 2002), 80.

159 **"the smoke . . . ":** Ibid., 79–80.

159 **The same railroads and steamships:** Ibid., 57–60.

159 **"It is not at all probable . . . ":** Ibid., 68.

161 **citizen-science project:** G. Sims and K. Russo, "Citizen Science on the Faroe Islands in Advance of an Eclipse," *Journal of the Royal Astronomical Society of Canada* 108, no. 6 (2014): 228.

Chapter 7: The Great American Eclipse and Beyond

172 **Australia is currently in the midst:** J. Tilley, personal communication, 2015.

175 **9.4 million visitors:** National Park Service, "Park Statistics," Great Smokey Mountains National Park website, www.nps.gov /grsm/learn/management/statistics.htm.

Chapter 8: The Last Total Eclipse

193 **"all attempts at imposition . . . ":** F. Baily, "On the Solar Eclipse Which Is Said to Have Been Predicted by Thales," *Philosophical Transactions of the Royal Society of London* 1 (1811): 236.

193 **determine the exact positions:** E. Halley, "Some Account of the Ancient State of the City of Palmyra, with Short Remarks upon the Inscriptions Found There," *Philosophical Transactions (1683–1775)* 19 (1695–1697): 175.

194 **"The real test of progress . . . ":** M. Stanley, "Predicting the Past: Ancient Eclipses and Airy, Newcomb, and Huxley on the Authority of Science," *Isis* 103, no. 2 (2012): 263–270.

194 **"there are tens of thousands . . . ":** S. Newcomb, *Reminiscences of an Astronomer* (New York: Houghton Mifflin, 1903), 64.

195 **unflinchingly dedicated:** Stanley, "Predicting the Past," 263.

195 **During the mania:** S. Newcomb, "Note on Accounting for the Secular Variations of the Orbits of Venus and Mercury," *Astronomical Journal* 14 (1894): 117–118.

195 **"Halley, notwithstanding his scientific merits . . . ":** S. Newcomb, "Researches on the Motion of the Moon" (Washington, DC: Government Printing Office, 1878), 257.

195 **new calculations derived:** Stanley, "Predicting the Past," 267.

196 **Halley had discovered:** F. R. Stephenson, L. V. Morrison, and G. J. Whitrow, "Long-Term Changes in the Rotation of the Earth: 700 B.C. to A.D. 1980," *Philosophical Transactions of the Royal Society of London*, series A, *Mathematical and Physical Sciences* 313, no. 1524 (1984): 48–53.

197 **60 million years ago:** G. H. Darwin, "On the Precession of a Viscous Spheroid, and on the Remote History of the Earth," *Proceedings of the Royal Society of London* 28 (1878): 193–194.

197 **The Pacific Ocean:** G. H. Darwin, "Problems Connected with the Tides of a Viscous Spheroid," *Proceedings of the Royal Society of London* 28 (1878): 196.

197 **no plausible mechanism:** J. A. O'Keefe and H. C. Urey, "The Deficiency of Siderophile Elements in the Moon," *Philosophical Transactions* 285, no. 1327 (1977): 569–575.

198 **how rare they might be:** H. C. Urey, "Chemical Evidence Relative to the Origin of the Solar System," *Monthly Notices of the Royal Astronomical Society* 131 (1964): 218–219.

198 **Both of these conclusions:** S. G. Brush, "Nickel for Your Thoughts: Urey and the Origin of the Moon," *Science* 217 (1982): 894–895.

198 **Since oxygen isotope ratios:** Ibid., 895.

199 **written extensively:** J. W. Truran and A. G. W. Cameron, "Evolutionary Models of Nucleosynthesis in the Galaxy," *Astrophysics and Space Science* 14, no. 1 (1971): 179–222.

199 **He and Ward proposed:** A. G. W. Cameron and W. R. Ward, "The Origin of the Moon," *Abstracts of the Lunar and Planetary Science Conference* 7 (1976): 120.

200 **new planet that was created:** R. M. Canup, "Forming a Moon with an Earth-Like Composition via a Giant Impact," *Science* 338 (2012): 1052–1054.

201 **same hot molten mix:** Ibid., 1054.

201 **recent computer models:** K. Pahlevan and A. Morbidelli, "Collisionless Encounters and the Origin of the Lunar Inclination," *Nature* 527 (2015): 492–493.

201 **"Had such a population . . . ":** R. Canup, "The Moon's Tilt for Gold," *Nature* 527 (2015): 456.

202 **as the bodies cooled and tidal friction grew:** B. A. Kagan and N. B. Maslova, "A Stochastic Model of the Earth-Moon Tidal Evolution Accounting for Cyclic Variations of Resonant Properties of the Ocean: An Asymptotic Solution," *Earth, Moon, and Planets* 66, no. 2 (1994): 173–175.

203 **over its lifespan:** G. E. Williams, "Geological Constraints on the Precambrian History of Earth's Rotation and the Moon's Orbit," *Reviews of Geophysics* 38, no. 1 (2000): 50.

203 **At its current, precisely known rate:** K. Lambeck, *The Earth's Variable Rotation: Geophysical Causes and Consequences* (New York: Cambridge University Press, 1980), 449.

203 **ancient seafloor sediments:** B. G. Bills and R. D. Ray, "Lunar Orbital Evolution: A Synthesis of Recent Results," *Geophysical Research Letters* 26, no. 19 (1999): 3045–3048.

203 **"whenever the subject becomes . . . ":** S. Newcomb, "The Course of Nature," paper presented at the American Association for the Advancement of Science, St. Louis, Missouri, 1878, 23.

204 **"primordial soup":** S. L. Miller and H. C. Urey, "Organic Compound Synthesis on the Primitive Earth," *Science* 130, no. 3370 (1959).

205 **Mars's obliquity:** J. Laskar et al., "Long Term Evolution and Chaotic Diffusion of the Insolation Quantities of Mars," *Icarus* 170, no. 2 (2004): 343–364.

205 **Moon's tidal embrace:** J. Laskar, F. Joutel, and P. Robutel, "Stabilization of the Earth's Obliquity by the Moon," *Nature* 361, no. 6413 (1993): 615–617.

205 **10,000 years ago:** J. D. Hays, J. Imbrie, and N. J. Shackleton, "Variations in the Earth's Orbit: Pacemaker of the Ice Ages," *Science* 194, no. 4270 (1976): 1130.

205 **Intelligent life on Earth:** P. Ward and D. Brownlee, *Rare Earth: Why Complex Life Is Uncommon in the Universe* (New York: Copernicus Books, 2000).

Index

Africa, 154–155, 157, 186
Africa Association, 90
agriculture, 23
airplanes, and eclipse-chasers,
    153–154
Alexandria, in Egypt, 46
Algeria, 94
Alhazen (aka Haytham, Ibn al-),
    46–50, 51, 77–79
Almagest (Ptolemy), 46
Anasazi. *See* Puebloans
Anaxagoras, 57–59, 69, 81–82
ancient buildings, 32
annular solar eclipse(s), 1 (photo), 11
    (fig.), 52, 191, 202
    of 478 BCE (2/1/478 BCE), 58–59
    of 1836, 91
    of 2012, 6 (photo)
    of 2013 (8/20/2013), 2 (photo)
    of 2017–2030, worldwide, 185
    (table)
    definition of, 11 (fig.)
    See also eclipses
Antarctica, 184
Antikythera Device, 63–64
Apollo mission, 197, 198–199
    and moon rocks, 198
Arabic influence, 49–50
Argentina, 10
Aristarchus, 69–70

Aristotle, 50, 61, 85
    and geocentric (or Earth-centered)
    model of the universe, 44–46
artistic skill, and transitory experi-
    ence of astronomer vs. camera,
    96–97
Asia, 186
astrology, 15–16, 42
    vs. astronomy, 18–19
    and constellations, 34–35
    and horoscopes, 35
    primal appeal of, 18
    and religion and science, 22–24
astronomers, the first, 19–22
Astronomical Observatory of
    Córdoba, 130n
Astronomical Unit (AU), 72, 73
astronomy, 17–19
    vs. astrology, 18–19
    and eclipses, transits, and
    occultations, 59–60
    need for, 19
    and photography, 94–96, 96 (fig.)
atoms, 98–99, 100–106
AU. *See* Astronomical Unit
Aubrey holes, 37, 105
Australia, 155, 160–161, 164, 172,
    186
Australopithecus, 19–20
Aztecs, 10–11

Babylon/Babylonians, 39, 40, 46
Baily, Francis, 89–93, 93–94, 172, 193
Baily's beads, 1 (photo), 9, 91–92,
    176–177
Baker Island, 172
Banks, Sir Joseph, 74
base-20 counting system of bars and
    dots. See Mayan base-20 count-
    ing system of bars and dots
Basra, 47
bematists, 60
bending of starlight near the Sun,
    128–130, 129 (fig.), 133–139
Bethe, Hans, 106
biological evolution, 22
black holes, 52–53, 139
blindness, warnings of, 13–15, 24–25.
    See also guide to safely viewing
    a solar eclipse
blood moon, 2
bone artifacts, 21
The Book of Optics (Alhazen), 48, 50
Boston, Massachusetts, 16–17
Bradley, George Young, 67
Bunsen, Robert, 98
burial positions, 21–22, 23

calendars, accuracy of, 30
California, 89n
Cambridge University, 133
camera, vs. artistic skill and transitory
    experience of astronomer, 96–97
camera obscura (a pinhole camera),
    7, 49
Cameron, Alastair, 199
Canada, 68, 172, 183
Canup, Robin, 201
carbon, 98, 105–106
Chacoans. See Puebloans
Chaldeans, 39–50
Chandrasekhar, Subrahmanyan, 133
Charon (moon of Pluto), 192
chiaroscuro, 87
Chile, 184
China, 9, 89

and court astrologers, 42
and solar eclipses, definition of,
    15–16
Christian tradition, 16, 43. See also
    religion
chromosphere, 104, 176
citizen-science, 150, 161–162, 195
Clark, William, 66, 68
climate, and tidal forces, 204–205
climate change, 205n
clocks, 64, 126
    of the Middle Ages, 44
    pendulum clock, 149
cognitive fluidity, 21
Cold War, 158
Columbus, Christopher, 17, 64–66
    and lunar eclipse, 65–66
Comet Swift-Tuttle, 116
comets, 192, 192n
    and impacts, 24
Comte, Auguste, 97, 98, 99
Concorde, 154–155, 155–156
Connecticut, 154
Consolmagno, Brother Guy, 206
constellations, 34–35, 71
continental drift, 197
Cook, James, 74
Copernicus, Nicolaus, 50–51, 78,
    120, 121
Cornell University, 150
corona, 1 (photo), 3 (photo), 9, 92,
    104, 176
    measurement of, 93–96
    photographs of, 93
    red flames of, as part of Sun or
        Moon?, 93–96
    temperature of, 104–105
coronaphiles, 144, 149, 155, 163,
    202–203
cosmos, size of, 69
Cottingham, Mr., 135
Course de la Philosophie Positive (Posi-
    tive Philosophy) (Comte), 97
court astrologers, Chinese, 42. See
    also Astrology

Cronkite, Walter, 171
Crucifixion, 16
cruise ships, and eclipse-chasers, 155,
    156–157
crystal spheres, universe of, 50
Curtis, Heber, 132

Dakota Territory, 17
dark matter, 53
Darwin, Charles, 101, 197
Darwin, George H., 197
De la Rue, Warren, 94–96, 96 (fig.)
Deimos (moon of Mars), 192
demons and gods, and eclipses, 204
Denmark, 147
Des Moines State Register, 115
Descartes, René, 77
diamond ring, 5 (photo), 151–153
distance
    around Earth, 62
    measurement of, 60
distance ladder out into the universe,
    79
Dresden Codex. See Mayan Dresden
    Codex
Drummond, Henry, 23
Duncan, Doug, 155, 156
Dyson, Frank, 134, 135–136, 139

Earth
    and close flybys and occasional
        impacts, 201
    days, years, and seasons on, 30, 32n
    formation of, 199–201
    and Heaven, 85–86
    measurements of (see under
        measurement)
    and Moon, and common event in
        formation of, 198, 199–201
    and Moon, and elemental iso-
        topes, 198, 198n
    and the Moon, contact between,
        and split in two (fission hypoth-
        esis), 197–198, 199
    and obliquity, 205

orbits the Sun vs. Sun orbits
    Earth, 70–71
    size of, beliefs about, 64
    slowing down of, 196
    slowing down of, and length of a
        day, 202, 203
    and Sun, distance between, 72
Earth 2.0 (planet), 80
Earth-centered (geocentric) model of
    the universe, 44–46
East India Company, 90
"eclipse," origin and meaning of, 9
eclipse calculator, 63
eclipse-chasers, 144, 145, 146,
    148–149, 158–159, 164
    and airplanes, 153–154
    commercial, 154–157
    and cruise ships, 155, 156–157
eclipse glasses. See under guide for
    safely viewing a solar eclipse
eclipse season, 36
eclipse "stories," 16–17
eclipse(s), 59–60
    dangers of, 24–25
    and demons and gods, 204
    frequency of, and moons, 67
    and knowable universe, 206–207
    and knowledge vs. feeling, 9
    and line of nodes, 35–36, 35 (fig.)
    and momentous events, associa-
        tion with, 16–17
    myths and rituals to explain, 3
    from omens to moments of awe,
        3, 25
    paths of, 11–12
    See also annular solar eclipses; hy-
        brid eclipses; lunar eclipses; total
        lunar eclipses; solar eclipses;
        total solar eclipses; transits
ecliptic, 34–36
Eddington, Arthur, 102, 107, 132–139
Egypt, 41, 48, 184, 186, 193
Eiffel, Gustave, 113
Einstein, Albert, 123–142
Eklund, Bárður, 157

electric fields, 122–123, 122n
electricity, 122, 125, 128
elemental isotopes, and Earth and
    Moon, 198, 198n
end-times stories, 16
England, 172
Equivalence Principle, 126
Eratosthenes of Cyrene, 60–63, 64,
    69–70
Europa (Jupiter's moon), 204
Europe, 5 (insert), 172, 186
    and clocks, of the Middle Ages, 44
European Southern Observatory, 184
evolution of species, 101
Exeligmos cycle, 39, 64
exoplanets, transiting, 80
expeditions. See solar eclipse expedi-
    tions; transit expeditions
experimentation, 118–120, 122
    and observation, and hypothesis,
    194

Faroe Islands, 6 (insert), 143–149,
    154, 156, 157–158, 159, 161–162,
    163–168, 166 (fig.)
feeling vs. knowledge, and eclipses, 9
Ferdinand (archduke of Austria),
    130
Ferrer, José Joaquín, 93
fission hypothesis, 197–198, 199
framework, 119–121
France, 76, 77–78, 162
Franco-Prussian War, 195
Franklin, Benjamin, 51
French Academy of Sciences, 74, 76
Freundlich, Erwin, 130
fuel, and reproducing nuclear fusion
    in the Sun, 107
Full Moons, and supermoons, 52

Galilei, Galileo, 46, 67–68, 72, 92,
    93, 103, 120, 121, 123, 206
    and telescope on the Moon,
    86–88, 87n
    and telescope on the Sun, 88

gases, of Sun/stars, 97, 98–100,
    102–107, 103n
general theory of relativity, 126–128,
    138. See also theory of relativity
geocentric (or Earth-centered) model
    of the universe, 44–46
geographic alignments, 32
German theory of relativity, 132
Germany, 76
glasses, eclipse. See under guide for
    safely viewing a solar eclipse
God, proof of, and knowable uni-
    verse, 206
God of the Gaps, 23, 203–204
gods and demons, and eclipses, 204
gravitational waves of spacetime, and
    black holes, 139
gravity, 102, 122, 126–127
    and bending of starlight near the
    Sun, 128–130, 129 (fig.), 133–139
    See also Newton's law of gravity
Great American Eclipse of 2017. See
    total solar eclipse: of 2017
Great Britain, 37
Great Houses, of Puebloans, 13, 34
Greek Islands, 58–59, 156
Greek mythology, 85, 106, 200
greenhouse gases, 107
Greenwich Observatory, 134, 139
guide to safely viewing a solar
    eclipse, 177–181
    and pinhole projectors, 178–179,
    180 (fig.)
    and safety glasses, 7, 14, 178–179
    what not to use, 179
    See also blindness, warnings of

Hall, Asaph, 118
Halley, Edmond, 72–74, 74–75, 149,
    172, 192–193, 203
    and the Earth, slowing down of,
    196
    and the Moon, acceleration of,
    195–196
Halley's Comet, 192n

Harriot, Thomas, 87n
Haytham, Ibn al-. *See* Alhazen
HD209458 (star), 80
Heaven, and Earth, 85–86
Heel Stone, 32
heliocentric model of solar system,
    50–51, 53
helium, 100, 102, 105, 106, 191
Hensley, Major, 153
Herodotus, 41, 193
Herschel, William, 97, 111
Himalayas, 94
Hipparchus, 62–63
Hippolytus of Rome, 59
historical records, and solar eclipses,
    and predictions, 193–196
Histories (Herodotus), 41, 193
Hogg, Helen Sawyer, 76
Holland, 76
Holm, Niels Pauli, 147
Homo erectus, 20
Homo habilis, 20
Homo sapiens, 20
Hopi (Native American tribe,
    American Southwest), 34
horoscopes, 35
Hoyle, Fred, 37, 105
Hungary, 5 (insert), 38, 162
hybrid eclipses, 185 (table). *See also*
    eclipses
hydrogen, 98, 99–100, 102, 103–104,
    105, 106–107
hypothesis, 118–119, 120, 125
    and observations and experimen-
    tation, 194

ice ages, 205
Illinois, 12n, 175, 183
immeasurability, 57
immensity, 57
India, 10, 89
Indonesia, 13–14
infrared, 97
intelligence life, and moons, 204,
    205–206

Intelligent Design, 22
Interstellar (film), 126
Ionians, 41–42
Iraq, 47, 193
iron, 104, 105–106
Irving, Washington, 64
Islamic calendar, 43
Italy, 43–44, 76, 92

Jamaica, 1–2, 3
Janssen, Pierre Jules, 99
Jefferson, Thomas, 66, 68
Jewish tradition, 43
Jupiter, 76–77
    moons around, and mapping,
    67–69, 77–78
    moons of, 191, 204
    and obliquity, 205

Kansas, 174
Kashi Khanda text, 89
Keeler, Ward, 13
Kentucky, 175
Kepler, Johannes, 50–53, 206
Kepler space telescope, 80–81
Kepler's laws of planetary motion,
    51–53, 73, 88, 121, 127
Kirchoff, Gustav, 98
knowable universe
    and eclipses, 206–207
    and proof of God, 206
knowledge vs. feeling, and eclipses, 9
Kuhn, Thomas, 119–120
Kuiper Belt, 77

La Science et l'hypothèse (*Science and
    Hypothesis*) (Poincaré), 125
La Silla Observatory, 184
Lahaina Noon, 61
law(s), 121–124, 122n
    vs. theory, 128
    See also Kepler's laws; Newton's
    laws
Le Gentil, Guillaume Joseph Hyacin-
    the Jean-Baptiste, 74–76

Le Verrier, Urbain-Jean-Joseph,
    112–114, 127
Leonardo da Vinci, 87
Lescarbault, Edmond Modeste, 114
Lewis, Meriwether, 66, 68
Lick Observatory, 130, 132, 139
light, 125, 128
    and Maxwell's equations, 122–124
light-years, 79
line of nodes, 35–36, 35 (fig.)
Lockyer, Norman, 99–100, 99n, 159
London, 148–149, 193
longitude, 65, 66
    and lunar eclipses, 62–63
    and occultations, 67–68
    and totality, 39
    see also measurement
Louis XIV, 68
Louville, le Chevalier de, 148–149
lunar eclipse(s), 4 (photo), 11 (fig.)
    of 1400s, 65–66
    of 2004 (10/27/2004), 16–17
    definition of, 11 (fig.)
    and measurement, 62–65, 66
    occurrences of, 30–31, 36
    partial, 4 (photo)
    and solar eclipses, pairing be-
        tween, 36
    See also eclipses
lunar tides, 196–198
lunation, 33–34, 36, 38n
Lydians, 41

MacKenzie, Alexander, 68
magnetic fields, 103–104, 122, 122n,
    123
magnetic monopole, 122n
magnetism, 122, 125, 128
Maine, 91–92
Makepeace, David, 164–165
Manhattan Project, 106
Manhattanhenge, 33
mapping
    of Canada, 68
    and eclipses, 66–69, 77–78

    of France, 77–78
    and Jupiter, moons around, 67–69,
        77–78
    of North America, 66
Mars (aka the Red Planet), 29–30,
    45, 51, 198
    annular eclipse of Sun by Mars's
        moon, 2 (photo)
    days, years, and seasons on, 30, 32n
    moons of, 30, 53, 118, 192,
        204–205
    obliquity of, and seasons, 204–205
    solar eclipse on, 53
MarsDial, 2 (photo), 29, 53
matter, 128
Maxwell, James Clerk, 122
Maxwell's equations, 122–124
Mayan base-20 counting system of
    bars and dots, 2 (photo), 37
Mayan Dresden Codex, 2 (photo), 37
measurement, 57–82
    of corona, 93–96
    and cosmos, size of, 69
    of distance, 60
    and distance ladder out into the
        universe, 79
    and Earth, curvature of, 61
    and Earth, distance around, 62
    and Earth, shape of, 60–61
    and Earth, size of, 64–66
    and Earth and Sun, distance
        between, 72
    and Earth orbits the Sun vs. Sun
        orbits Earth, 70–71
    and eclipse calculator, 63–64
    and eclipses, and mapping of
        North America, 66
    and eclipses and triangulation, 69
    and Jupiter, moons around, and
        mapping, 67–69
    and Kepler space telescope, 80–81
    and light-years, 79
    and longitude, 39, 62–63, 65, 66,
        67–69
    and lunar eclipses, 62–65, 66

and mapping of France, 77–78
and Mediterranean, longitude
    around, 62
and Moon, 58–59, 88
and Moon and Sun, size and dis-
    tance of, 69–70, 70 (fig.), 71 (fig.)
and parallax, 72, 78–79, 121
and planets, size and distance of,
    72–77
and planets, technique for discov-
    ery of, 79–80
and shadow size, 58–59
and solar eclipses, and finding
    locations, 66–67
and solar system, size of, 71–72
and speed of light, 77–79
and stars, distance to, 78–79
and summer solstice, 60–61
and Sun, distance to, 78, 79
and Sun, size of, 58–59
and Sun and Moon, size and dis-
    tance of, 69–70, 70 (fig.), 71 (fig.)
during total solar eclipse, and
    photographs, 94–96, 96 (fig.)
and transiting planets, 80
and transits, 59–60, 72–77, 80–81
and triangulation, 69
and the universe, 59–60
and Venus, distance to, 79
by walking, 60
Medes, 41
Mediterranean, 62, 156
Mercury, 120
    orbit of, 113, 117, 118, 119, 122,
        127–128, 132, 137, 140
    transits of, 113–114
Mexico, 164–165, 183
Michelson, A. A., 121–122, 123
Milankovitch cycle of climactic
    change, 205n
Miletus of Thales, 149
Milky Way, 19, 36, 184, 186
Miracle Card, 23
Missouri, 153n, 174–175
Mithen, Steven, 21

Montana, 173
Moon (Earth's), 85, 191–206
    acceleration of, 195–197
    and creationists, 204, 204n
    and Earth, and common event in
        formation of, 198, 199–201
    and Earth, and elemental isotopes,
        198, 198n
    and the Earth, contact between,
        and split in two (fission hypoth-
        esis), 197–198, 199
    features of, 86–88, 87n
    and lunar tides, 196–198
    measurement of, 58–59, 88
    movement of, 33–34, 34n
    or Sun, corona's red flames as part
        of, 93–96
    orbit of, 31, 201
    orbit of, and length of a day, 202,
        203
    orbit of, and solar eclipses, 192–193
    orbit of the, 149
    origin hypotheses, 197–202
    recession of, 197, 202–203, 204n
    and size and distance, measure-
        ment of, 69–70, 70 (fig.), 71 (fig.)
    size and orbit of, 150–151
    size of, 58–59
    and the telescope, 86–88, 87n
    and tidal forces, and climate,
        204–205
    uniqueness of, 191
moon rocks, 198
moon(s), 191–192, 204–205
    and intelligent life, 204, 205–206
    of Jupiter, 67–69, 77–78
    of Mars, 30, 118, 191
    of Saturn, 191
Morocco, 186
Mortensen, Lis, 146–147
motion, 128, 128. See also Kepler's
    laws of planetary motion; New-
    ton's laws of motion
myths and rituals, and eclipses, 3

NASA. *See* National Aeronautics
    and Space Administration
National Aeronautics and Space
    Administration (NASA), 29,
    53, 77, 80
  mission to Europa (Jupiter's
    moon), 204
National Science Foundation survey
    of 2014, 18, 45
NATO. *See* North Atlantic Treaty
    Organization
Nebraska, 174
Neptune, 112–113, 118
New Horizons spacecraft, 77
New Mexico, 4 (insert)
New York, 92
New York City, 149–154
New York City grid, 33
New York Times, 115, 117, 137–138,
    150, 152, 153
Newcomb, Simon, 194–195, 203
Newton, Isaac, 88, 97, 111, 121, 126,
    135, 136
Newton's law of gravity, 88, 112, 118,
    127, 128, 137, 192, 194
Newton's laws of motion, 149
Niagara Falls, 183
nodes, line of, 35–36, 35 (fig.)
Norsemen/Norse tradition, 10, 16,
    143–144
North America, 66
North Atlantic Treaty Organization
    (NATO), 158
North Carolina, 175
Norway, 144
Nubia, 21–22, 23
nuclear bombs, 106
nuclear fusion, 101–102, 105, 106–107

obliquity, and seasons, 204–205
observation, 31
  and hypothesis, and experimenta-
    tion, 194
observatories
  accidental, 32–33

  See also specific observatories
Observatory, 135
Occam's Razor, 119
occultations, 59–60
Ojibwe (Native American tribe), 10
*On the Sizes and Distances of the Sun
    and the Moon* (Aristarchus), 69
optics, 122
  and vision, laws of, 49
orbit(s)
  calculation of, 192
  Earth vs. Sun, 70–71
  of Mercury, 113, 117, 118, 119, 122,
    127–128, 132, 137, 140
  of Moon, 31, 149, 150–151, 201
  of Moon, and length of a day, 202,
    203
  of Moon, and solar eclipses,
    150–151, 192–193
Oregon, 14–15, 173
Origin of Species (Darwin), 101
oxygen, 98, 105–106, 198

Pandora's Box, 106, 107
Pang, Alex Soojung-Kim, 158–159
paradigms, 119, 120–121
Paraguay, 10
parallax, 72, 78–79, 121
Paris Observatory, 112, 195
Peloponnese, 58–59
pendulum clock, 149
Perrine, C. D., 130n
Perseid meteor shower, 116, 186
Persian Gulf, 39, 41
Peru, 94
petroglyph(s), 32
  representing total solar eclipse, 4
    (photo)
Phobos (moon of Mars), 53, 192
  annular eclipse of the Sun by
    Mars's moon, 2 (photo)
photographs/photography
  and astronomy, 94–96, 96 (fig.)
  and measurement during total
    solar eclipse, 94–96, 96 (fig.)

recommendation not to take, 181
of solar eclipses, 181–183
photosphere (surface of light), 88, 99, 102–103, 104
physical law, 100
physics, 48
Piazza San Marco (in Venice), 43–44
Picasso, Pablo, 125
pictographs, 12
Pike, Zebulon, 68–69
Pike's Peak, 69
pinhole projector. See under guide for safely viewing a solar eclipse
Plains Indians, 17
planetary motion. See Kepler's laws of planetary motion
planetesimals, 199
planet(s), 85
  discovery of, 79–82, 111–113
  formation, 199
  measurement of size and distance of, 72–77
  number of, 79–80
  obliquity of, and seasons, 204–205
  suitable for life, 81–82
  technique for discovery of, 79–80
  transiting, 80
  See also specific planets
Pliny the Elder, 41–42
Plutarch, 59
Pluto, 77, 192
Poincaré, Henri, 124–125
Point Venus (Tahiti), 74
Pomo (Native American tribe, Northern California), 36
Powell, John Wesley, 66–67
predictions, of solar eclipses, 3, 17–18, 149, 203
  and historical records, 193–196
  throughout history, 29–53
Princeton, 150
principle of relativity, 123–124, 126
Prussian Academy of Sciences, 127
Ptolemaeus. See Ptolemy

Ptolemy (aka Ptolemaeus), 46, 49, 50, 51, 120, 121
Puebloans (ancient Native American culture; aka Anasazi; Chacoans), 4 (photo), 12, 13, 34

racism, 158–159
radiation, 107
radioactivity, 101
radiometric dating, 197
Ragnarok, 16
Raleigh, Sir Walter, 66
Rapture, 16
Red Planet. See Mars
red prominences. See solar prominences
Red Sox World Series "curse," 16–17
refracting telescope, 66
Refutation of All Heresies (Hippolytus of Rome), 59
religion, 22–24, 206. See also Christian tradition
religious holidays, 42–43
representational art, 21
ring of fire, 11 (fig.)
rituals and myths, and eclipses, 3
robotic explorers, 29–30
Roddenberry, Gene, 128n
Rømer, Ole, 77–78
rovers, on Mars, 2 (photo), 29–30, 53
Royal Astronomical Society, 90–91, 136
Royal Society, 136
Russell, H. C., 96–97
Russia, 76
Russo, Kate, 146, 157–158, 160–165

Sagan, Carl, 106, 204
Saros cycle, 38–39, 38n, 41, 43, 64
Saros Cycle Texts, 40
Saros Siblings, 162–163
Saturn, moons of, 191
Scandinavia, 10
science, and astrology and religion, 22–24

Scotland, 91
seasons, and obliquity, 204–205
shadow
    size, and measurement, 58–59
    of solar eclipse, 31
shadow size, and measurement, 58–59
Simon, Carly, 155
Sims, Geoff, 161, 163–164
Sioux City Daily times, 115
Sioux (Native American tribe,
    Plains Indians), 17
61 Cygni, 78–79
Sobel, Dava, 68
Solar Dynamic Observatory space-
    craft, 4 (insert)
solar eclipse expeditions, 66–69, 91,
    94–96, 133–135, 139, 158–159
    equipment list, 95
solar eclipse(s), 11 (fig.), 89–93
    of 400 BCE, 58
    of 585 BCE (5/28/585 BCE), 41
    of 603 BCE (5/18/603 BCE), 41
    of 1781, 111
    of 1886, 118
    of 1922, 139
    of 1983 6/11/1983, 13–14
    and bending of starlight near the
        Sun, 128–130, 129 (fig.), 133–139
    and blindness (*see* blindness,
        warnings of)
    definitions of, 11 (fig.), 15–16
    description of, 7–9
    expeditions (*see* solar eclipse
        expeditions)
    fearful reactions to, 9–11, 12–13
    length of, 13
    and lunar eclipses, pairing
        between, 36
    on Mars, 53
    and measurement (*see under*
        measurement)
    and Moon, orbit of, 192–193
    occurrences of, 30, 36
    predictions (*see* predictions, of
        solar eclipses)

    shadow of, 31
    viewing of (*see* guide to safely
        viewing a solar eclipse)
    See also eclipses
solar flare, 3 (photo)
solar prominences (geysers of
    hydrogen gas erupting off Sun's
    surface), 1 (photo), 3 (photo),
    99, 103–104, 176–177
solar system(s)
    extent of, 72–82
    formation of, 198
    heliocentric model of, 50–51, 53
    measurement of, 71–72
    number of, 79
solar telescope, 95
Sørensen, Súsanna, 167–168
Sousa, John Philip, 76
South Carolina, 176
South Pacific, 94
space and time, 123, 124–125, 128
Spain, 95, 186
special theory of relativity, 125. *See
    also* theory of relativity
spectral lines, 98, 104, 105–106
    instruments to study, 99–100
    and solar prominences, 99
speed of light, 77–79
Star Trek (film; television series), 128n
starlight (near the sun), bending of,
    128–130, 129 (fig.), 133–139
stars, 85
    collapse of, 105–106
    composition of, 97, 98–100,
        102–107
    distance to, 78–79
    element composition of, 98–106,
        107
    formation, 199
    life cycle of, 100–105
    names of, and Arabic influence,
        49–50
    and nuclear fusion, 106–107
    as part of us, 106
    temperature of, 104–105

Stonehenge, 32, 33–34, 99n
  and Aubrey holes, 37, 105
  and Heel Stone, 32
The Structure of Scientific Revolutions
  (Kuhn), 119
summer solstice, 60–61
Sun, 85–86, 88–107
  and atoms, 98–99, 100–106
  and Bailey's beads, 91–92
  and chromosphere, 104
  collapse of, 105–106
  composition of, 97, 98–100,
    102–107
  and corona, 92, 93–96, 104–105
  distance to, 78, 79
  and Earth, distance between, 72
  element composition of, 98–106,
    107
  false color image of, 3 (photo)
  as featureless, 86
  and Galilei, Galileo, 88
  and gases, 97, 98–100, 102–107
  life cycle of, 100–105
  and measurement of size and
    distance, 69–70, 70 (fig.),
    71 (fig.)
  and nuclear fusion, 105, 106–107
  and obliquity, 205
  or Moon, corona's red flames as
    part of, 93–96
  orbits the Earth vs. Earth orbits
    the Sun, 70–71
  and photosphere (surface of light),
    88, 99, 102–103, 104
  size of, 58–59
  spacecraft image of, 4 (photo)
  and spectral lines, 98, 99–100, 104,
    105–106
  and sunlight, colors of rainbow
    within, 97–98
  and sunspots, 88–89, 89n, 103
  temperature of, 104–105
  at the zenith, 61
sun cycles, observation of, 31–32
sundials, 2 (photo), 29–30

sunlight, colors of rainbow within,
  97–98
sunspots, 3 (photo), 88–89, 89n, 103
  spacecraft image of, 4 (photo)
supermoons, 52
supernova explosions, 24, 105–106
Svalbard, 144
Swift, Lewis, 116–118
Syria, 193

Tahiti, 74
technology, in the nineteenth century, 113
telescope(s), 86–88, 87n, 121
  Kepler space, 80–81
  refracting, 66
  solar, 95
Teller, Edward, 106
Tennessee, 175
TESS. *See* Transiting Exoplanet
  Survey Satellite
Texas, 183
Thales of Miletus, 41–42, 58, 90, 193
Theia (hypothetical ancient planetary world), 200
theory
  vs. law, 128
  successful, 194
theory of gravitation, 132
theory of relativity, 123–142. *See also*
  general theory of relativity;
  special theory of relativity
thermal radiation, 102
tidal forces, and climate, 204–205
time and space, 123, 124–125, 128
Times (London), 135–136, 137
Torre dell'Orologio, 44
Total Addiction (Russo), 160
total lunar eclipse(s), 1–3, 1 (photo)
  of 1504 (2/29/1504), 1–2, 3
  of 2007 (8/2007), 186
  of 2017–2030, worldwide, 187
    (table)
  of 2018 (7/27/2018), 186
  length of, 2

total lunar eclipse(s) *(continued)*
  and totality, 186
  See also eclipses
total solar eclipse(s), 1 (photo), 11
    (fig.)
  of 610 BCE (9/30/610 BCE), 90
  of 1097 (7/11/1097), 4 (photo), 12
  of 1612 (5/30/1612), 143–144
  of 1715 (4/22/1715), 149, 172, 193,
    195
  of 1780, 91–92
  of 1806, 92
  of 1842 (7/18/1842), 3 (photo), 91,
    92, 172
  of 1860, 94–96, 96 (fig.)
  of 1869, 17
  of 1871, 159
  of 1914 (8/21/1914), 130, 131 (fig.)
  of 1918, 172
  of 1919 (5/29/1919), 133–139
  of 1925 (1/24/1925), 149–154, 152
    (fig.)
  of 1954 (6/30/1954), 147, 158
  of 1970s, 154–156
  of 1972, 155
  of 1973, 154–155, 155–156
  of 1974, 155
  of 1974 to 2038, 172
  of 1979 (2/26/1979), 14–15,
    171–172
  of 1991, 164–165
  of 1999 (8/11/1999), 5 (photo), 38,
    162–163
  of 2002, 164
  of 2006, 156
  of 2012, 6 (photo), 160
  of 2013 (11/3/2013), 5 (photo), 6
    (photo), 157
  of 2015 (3/20/2015), 6 (photo),
    144–149, 154, 156, 157–158, 159,
    161–162, 163–168, 166 (fig.)
  of 2015–2065, 8 (photo)
  of 2017 (8/21/2017), 6 (photo), 7
    (photo), 12n, 38, 80, 103n, 153n,
    172, 173–177

  of 2017–2030, worldwide, 185
    (table)
  of 2019 (7/2019), 184
  of 2021 (12/4/2021), 184
  of 2024 (4/8/2024), 12n, 103n, 172,
    183
  of 2027 (8/2/2027), 184, 186
  of 2028 (7/22/2028), 186
  of 2030 (11/25/2030), 186
  of 2044, 172
  of 2045, 172
  of 2052, 172
  and blindness (*see* blindness,
    warnings of)
  commemorative stamps of, 5
    (photo)
  commemorative travel posters of,
    6 (photo)
  community preparation for,
    160–162
  cycles, 172
  definition of, 11 (fig.)
  differences between, 15
  future, 183–186, 185 (table),
    187 (table)
  the last, 202–203
  measurement during, and photo-
    graphs (*see under* measurement;
    photographs/photography)
  and the Moon, size and orbit of,
    150–151
  occurrences of, 10–12, 12n, 39,
    201–202
  patterns of, 10–11
  petroglyph representing, 4 (photo)
  photographs (*see* photographs/
    photography)
  and starlight near the Sun, bending
    of, 128–130, 129 (fig.), 133–139
  types of, 11 (fig.)
  viewing of (*see* guide to safely
    viewing a solar eclipse)
  See also Baily's beads; corona;
    diamond ring; eclipses; solar
    prominences; totality; umbra

totality, 9, 145, 148, 176–177
    and longitude, 39
    map of, for eclipse of 8/21/2017,
        7 (photo)
    occurrences of, 12
    prolonging, in airplanes, 154–155
    and total lunar eclipses, 186
tourism, 158, 159
transit expeditions, 74–76
Transit of Venus March (Sousa),
    76
Transiting Exoplanet Survey
    Satellite (TESS), 80
transiting exoplanets, 80
transiting planets, 80
transitory experience, and artistic
    skill of astronomer, vs. camera,
    96–97
transit(s) (tiny eclipses), 59–60,
    72–77, 80–81, 128
    of Mercury, 113–114
    of Venus, 72–73, 76, 113
    of Vulcan (hypothetical planet),
        114–115, 117
    See also eclipses
triangulation, and measurement,
    69
trigonometry, 69–70, 70 (fig.)
Turin Observatory, 130
Turkey, 39, 41
Twain, Mark, 2–3
2001: A Space Odyssey (film), 126
Tyson, Neil deGrasse, 19, 33

ultraviolet, 97–98
umbra, 69, 176, 187 (table)
United States, 76, 172, 173–177,
    183

future total solar eclipses in, 12n,
    38, 103n (*see also* total solar
    eclipses: of 2017 and 2024)
    See also specific states
universe
    of crystal spheres, 50
    geocentric (or Earth-centered)
        model of the, 44–46
    and measurement, 59–60
University of Texas, 139
Uranus, 111–112
Urey, Harold, 198, 199, 204

Vatican Observatory, 206
Venus, 12, 72–74, 79, 118, 191–192
    transits of, 72–73, 76, 113
Verdet, Jean-Pierre, 10
Vulcan (hypothetical planet), 114–
    120, 122, 128, 128n, 132, 195
    transits of, 114–115, 117

walking, measurement by, 60
Ward, William, 199
Washington (state), 171–172
Watson, James Craig, 116–117
wavelength, 99
Williams, Rev. Samuel, 91–92
World Astronomy Heritage Site, 42
World Series "curse," and lunar
    eclipse, 16–17
World War I, 130
Wyoming, 174, 186

Yale, 150
Yale Observatory, 154
Young, C. A., 117

Zeiler, Michael, 145

Photo courtesy of Carrie Rosema

Tyler Nordgren is an astronomer and professor of physics at the University of Redlands. He has worked as an astronomer at both the US Naval Observatory and Lowell Observatory. Since 2007, Nordgren has worked with the National Park Service. He lives in Claremont, California.